W0111872

Fundamentals of
Pipeline Engineering

Fundamentals of Pipeline Engineering

Editor

Anjana Shrivastav

Fundamentals of Pipeline Engineering
Edited by **Anjana Shrivastav**

Printed in 2017

ISBN: 978-1-68117-379-5

Library of Congress Control Number: 2015941566

© 2016 by
SCITUS Academics LLC,
616, Corporate Way, Suite 2, 4766,
Valley Cottage, NY 10989

www.scitusacademics.com

This book contains information obtained from highly regarded resources. Copyright for individual articles remains with the authors as indicated. All chapters are distributed under the terms of the Creative Commons Attribution License, which permits unrestricted use, distribution, and reproduction in any medium, provided the original author and source are credited.

Notice

Reasonable efforts have been made to publish reliable data and views articulated in the chapters are those of the individual contributors, and not necessarily those of the editors or publishers. Editors or publishers are not responsible for the accuracy of the information in the published chapters or consequences of their use. The publisher believes no responsibility for any damage or grievance to the persons or property arising out of the use of any materials, instructions, methods or thoughts in the book. The editors and the publisher have attempted to trace the copyright holders of all material reproduced in this publication and apologize to copyright holders if permission has not been obtained. If any copyright holder has not been acknowledged, please write to us so we may rectify.

Contents

Preface

Pipelines perform vital functions. They serve as arteries, bringing life-dependent supplies such as water, petroleum products, and natural gas to consumers through a dense underground network of transmission and distribution lines. They also serve as veins, transporting life-threatening waste (sewage) generated by households and industries to waste treatment plants for processing via a dense network of sewers. Because most pipelines are buried underground or underwater, they are out of sight and out of mind of the general public. The public pays little attention to pipelines unless and until a water main leaks, a sewer is clogged, or a natural gas pipeline causes an accident. However, as our highways and streets become increasingly congested with automobiles, and as the technology of freight pipelines continues to improve, the public is beginning to realize the need to reduce the use of trucks and to shift more freight transport to underground pipelines. Pipeline engineering requires an understanding of a wide range of topics. Operators must take into account numerous pipeline codes and standards, calculation approaches, and reference materials in order to make accurate and informed decisions. Pipeline Engineering provides concise, easy-to-use, and accessible information on onshore and offshore pipeline engineering. Topics covered include: design; construction; testing; operation and maintenance; and decommissioning.

Editor

Chapter 1

Theoretical Evaluation of Two-Phase Flow in a Horizontal Duct with Leaks

Morgana de Vasconcellos Araújo[1],
Severino Rodrigues de Farias Neto[1], and
Antonio Gilson Barbosa de Lima[2]

[1]Department of Chemical Engineering, Center of Science and Technology, Federal University of Campina Grande (UFCG), Campina Grande, Brazil

[2]Department of Mechanical Engineering, Center of Science and Technology, Federal University of Campina Grande (UFCG), Campina Grande, Brazil

ABSTRACT

The transport of oil and its derivates are done, mostly, by pipeline. The time to detect leaks has to be short for preventing big disasters in the nature and decreasing losses for industries. The techniques available for leak detection vary from visual inspection to the use of computational techniques such as mathematical modeling. This paper aims to study the fluid dynamics of two-phase flow (water-oil) in the pipe with leakage. The equations of the mass and momentum conservation are numerically solved by using the ANSYS® CFX commercial code with the aid of a structured mesh of a horizontal pipe with three holes of leaks. The Eulerian-Eulerian model was adopted considering the oil as continuous phase and water as dispersed phase, and constant fluid properties. With profiles of pressure and volume fraction along the time in the pipe, the influence of leakage on the single-phase (oil) and two-phase (water-oil) was evaluated.

INTRODUCTION

The transportation of oil and its derivatives are mostly conducted through pipelines that connect production facilities, refineries and, in some situations, the consumer centers.

The materials used for making pipes come through technological improvements, where there is a significant use of materials based on special steel, lighter and stronger. However, despite these advances, there are still problems with leaks in pipelines generating great interest from the oil industry in view of the high costs incurred by financial services, potential risks and environmental costs.

Environmental disasters related to the oil spill in addition to degrading the environment, are responsible for spending millions of dollars in remediation.

In oil activities, particularly in transportation by pipelines, accidents have happened, causing financial and environmental losses.

There are currently a variety of techniques available for the detection of leaks, ranging from simple physical inspection by acoustic methods. Zhang [1] classified the detection methods into three categories: observation (perhaps the simplest and most ancient, is conducted through a visual inspection noting any ponding on the soil surface or anomalous growth of vegetation), direct detection (devices are used for the detection and location of leakage) and indirect detection (software is used based on mathematical models which allow to perform detection by means of data flow like pressure, temperature, mass flow rate, etc.).

The faster the identification of a leak in a pipeline, the faster valves are closed and the pumps will stop and, consequently, the greater the chances of avoiding a catastrophe are. However, in order to do the detection and precise identification of the position of a leak, it is necessary to know the behavior(s) of fluid(s) within the duct which allows determination of pressure drop between two points being evaluated.

According to Buiatti [2] the kind of leaks appearing in pipe networks can be divided into two classes:

- Leak by "breaking" the tube—occurs less frequently, but it is dangerous due to the amount of product spilled in the vicinity of the leak. However, these disruptions are easy to detect because they are accompanied by pressure losses and volumetric differences.

- Leakage of small proportions—Small leaks around 5 liters per hour are difficult to detect due to their sizes and can cause large losses of products to get noticed. Maybe provided by corrosion, fatigue of the material that makes up the pipe or by failure is in welds. There are few methods capable of detecting leaks of this order.

The occurrence of a leak surely results in a disturbance in the behavior of pressure in the pipe and this disturbance can describe the location, the size and the altitude of the leak. A dynamic associated with the leak is propagated along the pipe from the point of the leak and with a variable speed, being perceived in different moments by sensors on the pipe. The location of the leak in the pipe is calculated from the time interval between the perceptions of the dynamic data from sensors coupled to the distance between the sensors. Azevedo [3] proposed the division of the temporal evolution of a leak in three steps:

- Pre-Leak: it corresponds to the behavior of the flow in the pipe before the leak phenomenon and reflects the normal steady-flow conditions;
- Transitional: it corresponds to the behavior of accommodation of the flow in the pipe from the initial leak until the moment of reaching a new steady state;
- Post-Leak: it corresponds to the behavior of the flow in the pipe after the occurrence of a leak and the stabilization of the flow conditions, reflecting the flow stationary conditions on pipes with the presence of a leak Different works related to leakage in the pipe has been reported in the literature [4-8].

Agbakwuru [4] discusses about the challenges and technologies available for viewing and examination of leaks in pipelines located in the sea and under the conditions of poor visibility. It is proposed the use of Remotely Operated Vehicles with an optical eye as a tool to perform visual inspection of the pipeline and assist in the challenge of repairing the oil spill. The techniques used for pipelines submerged in water of poor visibility are based on tools used: hydrophone/Acoustic sensor, direct hydrocarbon leak detection, fluorescence. This author did a research about the oil/gas industry in the Nigeria Niger Delta Region. The causes of pipeline leaks were classified in operational, structural, unitended and intended damages. The Autor has shown the big values of leakes caused by sabotage in the Delta Region from the year 2005 to 2011. To reduce cost and improve effectiveness of systems of ducts in Delta Niger, it was proposed a combination of two methods of leak detection: the impressed alternating cycle current (IACC) for pipes with a small diameter (4 to 8 inches) and short length and the acoustic method for pipes with large diameter and long length.

Braga [5] evaluated air and water flow in a pipe with 1250 meters length. Coupled pipe flow is a data acquisition system, where this detection system has four pressure transducers connected to a computer equipped with an ADA converter board (Analog-digital-Analog). Three different types of flow regimes were studied: isolated bubbles of air without leak of the fluid, flow of a single air bubble in the presence of leak, and continuous flow air-water with leak. The experiments showed that the presence of air in the pipe generates reflections of waves (spring effect) that interferes with the detection of the leak. The author concluded that depending on the amount of air in the pipe, this can act as a damper of the shock wave, reducing the impact produced by leakage and thus, decreasing the sensitivity of the system.

The pressure transducers along the pipe convert pressure variations into variation of an electrical magnitude (voltage or current) to the PLC (programmable logic controller) which can work with this information. To drain compressed air Bezerra [6] built a pipe in PVC (polyvinyl chloride) with a diameter of 2.5 cm and a length of 5.56 meters. Two transducers were placed along the pipe and the transient pressure (pressure variation in the pipe with time) was measured simultaneously. The method detected the leak in the pipe, but we observed a difficulty in obtaining the location of the leak and this was explained by the fact that the two transients occur with a very small time difference, which was justified by the fact that the pipe had a short length. Bezerra [6] concluded that the program was successful in locating the leak which was necessary for the transducers installed at a distance farther than 12.24 meters from each other.

Garcia et al. [7] represented with the aid of commercial software COMSOL Multiphysics 4.0a®, a section of a pipe with 2 feet in length, 10 centimeters in diameter and a orifice leak of 4 mm in diameter. A cylindrical volume was linked perpendicularly to orifice surface to observe the leak jet. At the beginning, the pipe was completely filled with water at an absolute pressure of 101,325 Pa and the cylindrical volume had air at an absolute pressure of 0 Pa. Initially a leak test with a bench trial was held, where there was water inside the pipe and air was injected to study the flow behavior in the region of leakage. The numerical simulations were performed with the help of a mesh with 21.376 tetrahedral elements, with injections of air with a speed of 2 m/s and 10 m/s during a time interval of 1 second, considering laminar flow. This author analyzed the magnitude of the velocity profiles in a central plane of the section of duct and observed that in the region of the leak there was an increase in speed, and, in the simulation with the higher speed, the values in the leakage region reached 425.5 m/s featuring a supersonic behavior in this region.

Sousa et al. [8] simulated a rising flow of an isothermic water/oil two-phase mixture. The domain of flow is a vertical duct 8 meters long and a diameter of 15 cm, with an orifice leak of 6 mm located at 4 m of inlet section. The mesh was done on the domain and has 327,327 hexahedral elements. The simulation was performed in ANSYS CFX 11.0. The authors varied the volumetric fraction of oil in the inlet flow duct in amounts of 0.75 to 1 and it was noted that the smaller was the oil fraction, the major was the water fraction in the mixture and, consequently, the major was the mass flow through the orifice leak. The inlet velocities of mixture were varied and it was possible to realize that from 0.75 to 1.5 m/s the rate between the mass flow of leakage and total mass flow in the inlet section decreased with the increase of the velocity, but between 1.5 and 2 m/s the effect was the opposite due to inertial forces of flow being smaller than the forces caused by pressure drop from inlet to the leak. Complementing the authors' conclusions, it is observed that the emergence of the leak causes a pressure drop which is directly proportional to the volume fraction of water in the mixture and the flow rate of the same.

In this context, the proposed research is to contribute to the study of leakage and help to understand the phenomenon involved via techniques of computational fluid dynamics. This work aims to study numerically the twophase flow (water-oil) in a horizontal tube in the presence of leakage.

METHODOLOGY

Study Domain

To study the influence of leak in the hydrodynamics of the oil-water two-phase flow in a horizontal pipe, was adopted a study domain. The representation of the computational domain was done with the aid of ICEM-CFD 12.1 with points, curves and surfaces (**Figure 1**). The study area consists of a horizontal pipe with 10 meters long and 20 inches in diameter. The tube contains two holes of 1.6 cm diameter at the points (x = 5 m, y = 0.1 m, z = 0 m) and (x = 7.5 m, y = 0.1 m, z = 0 m) and a third hole of 3 cm diameter at the point (x = 5 m, y = −0.1 m, z = 0 m).

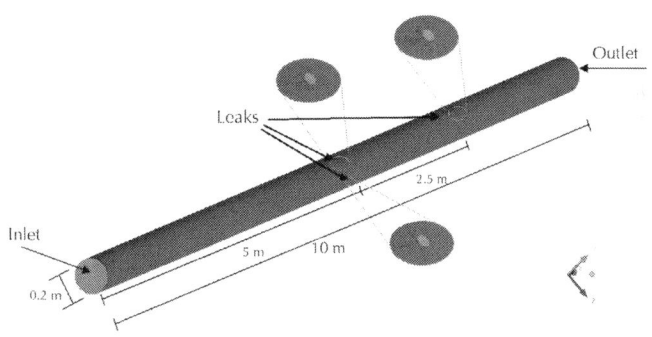

Figure 1: Geometry with the regions of leak.

The representative mesh of the pipe was generated using the block strategy, which a single block is initially created involving the area of study and then the block is subdivided into several others. This strategy allows greater control of mesh refinement in regions desirable. The mesh was refined on the geometry and contains approximately 327,000 control volumes as shown in Figure 2.

Mathematical Modeling

In the studied flows, both monophase (oil) and the twophase (water-oil) flows, they are based on the following considerations:

- Transient and laminar flow;
- The fluid is newtonian and incompressible and the physicochemical properties are constants;
- Isothermal process;
- There is no occurrence of chemical reactions;

- It was considered the gravitational effect;
- No transfer of mass and momentum at the interface water and oil phases;
- The interfacial forces of no drag (forces of lift, wall lubrication, virtual mass, pressure and turbulent dispersion of solid) were neglected.

The equations that describe the flow was withdraw from CFX 12.1®.

The equation of mass conservation is:

$$\frac{\partial(f_\alpha \rho_\alpha)}{\partial t} + \nabla \cdot (f_\alpha \rho_\alpha \bar{U}_\alpha) = 0 \tag{1}$$

where the the the greek subscript α represents the involved phase in two-phase mixture of water/oil; f, p, and $\bar{U}$ are respectively the volumetric fraction, density and velocity vector. For the phase α, the velocity vector is given by $\bar{U}\alpha = (u, v, w)$.

The equation of momentum is:

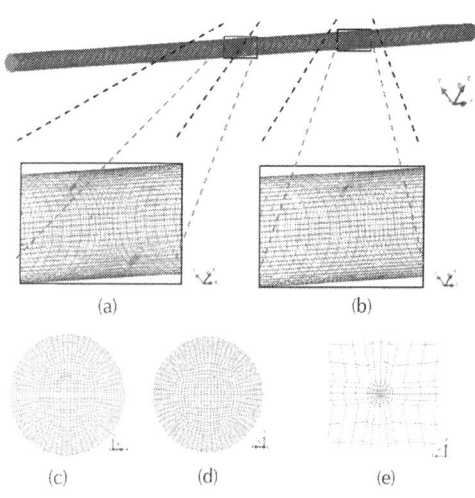

(a)   (b)

(c)   (d)   (e)

Figure 2: Representation of the mesh tube and the detail of section of (a) leaks at x = 5m; (b) leak at x = 7.5 m; (c) input; (d) output and (e) one of the leaks.

$$\frac{\partial\left(f_\alpha \rho_\alpha \vec{U}_\alpha\right)}{\partial t} + \nabla\cdot\left[f_\alpha\left(\rho_\alpha \vec{U}_\alpha \otimes \vec{U}_\alpha\right)\right] =$$

$$-f_\alpha \nabla p_\alpha + \nabla\cdot\left\{f_\alpha \mu_\alpha\left[\nabla\vec{U}_\alpha + \left(\nabla\vec{U}_\alpha\right)^T\right]\right\} + \vec{S}_{Ma} + \vec{M}_\alpha$$

(2)

where p is the pressure, $\vec{S}_{Ma}$ represents the external forces acting on the system per unit volume, $\vec{M}_a$ describes the overall force per unit volume (interfacial drag forces).

The interfacial drag force is given by:

$$\vec{M}_\alpha = \frac{3}{4}\frac{C_d}{d_p} f_\beta \rho_\alpha \left|\vec{U}_\beta - \vec{U}_\alpha\right|\left(\vec{U}_\beta - \vec{U}_\alpha\right)$$

(3)

where d_p is the particle diameter and C_d is the drag coefficient calculated using the Schiller-Neumann correlation, as follows:

$$C_d = \max\left[\frac{24}{Re}\left(1 + 0.15\,Re^{0.687}\right), 0.44\right]$$

(4)

It was adopted the model of particle (Eulerian-Eulerian approaches) where oil behaves as a continuous fluid and water as spherical particles dispersed.

The simulations occurred in transient regimes and time steps were applied with 15 iterations each. In a period 0 to 0.01 second the flow time step used was 0.001 second; from 0.01 to 0.1 second the time step value increased to 0.01 second and after elapsed time 0.01 second the time step became 0.1 second.

The initial and boundary conditions are:

    1) At the inlet:
        a) The monophase flow has the characteristics:
            i)   Speed of oil-u_x = 0.2 m/s;
            ii) Entrance length: Le = 1.43 m.
        b) For the two-phase flow (water-oil) we have:

      i. The speed of the mix flow of water and oil-u_x = 0.2 m/s;

      ii. Volume fraction-f_{oil} (**Table 1**);

2) At the output: prescribed static pressure equal to 101,325 Pa;.

3) At the wall:

    a)   No slip: $u_x = u_y = u_z = 0$;

    b)   The smooth tube;

    c)   The static walls.

    d)   In the leak: average static pressure with value 101,325 Pa.

4) Pressure of reference: 0.0 Pa.

Physico-Chemical Properties

The properties of the fluids used in the present work, are shown in Table 2.

Case Studies

The simulations were performed to analyze the influence of the oil volume fraction in the amount of oil leaked, as the disorder caused by the presence of two leaks in hydrodynamic flow.

Table 1: Data used in the simulations

Case	Volumetric fraction		Pressure at the leakage Sections [Pa]		
	Oil f_{oil}	Water f_{water}	Leak 1	Leak 2	Leak 3
1	1.000	0.000	101,325	-	-
2	1.000	0.000	101,325	101,325	-
3	0.990	0.005	101,325	-	-
4	0.990	0.005	101,325	101,325	-
5	0.975	0.025	101,325	-	-
6	0.975	0.025	101,325	101,325	-
7	0.950	0.050	101,325	-	-

8	0.950	0.050	101,325	101,325	-
9	0.925	0.075	101,325	-	-
10	0.925	0.075	101,325	101,325	-
11	0.900	0.100	101,325	-	-
12	0.900	0.100	101,325	101,325	-
13	1.000	0.000	-	-	101,325
14	0.950	0.050	-	-	101,325
15	0.900	0.100	-	-	101,325

Table 2: Fluid properties

Properties	Continuous phase (oil)*	Disperse phase (water)**
Density (kg/m³)	868.7	997
Molar mass (kg/kmol)	105.47	18.02
Viscosity (Pa·s)	0.17	0.0008899
Particle diameter (mm)	0.001	
Surface tension (N/m)	0.03	

*Source: Cunha [9]; **Source: Welty et al. [10].

The influence of orifice size of leak in the monophase flow of oil and two-phase flow of oil and water were also studied. The constructed geometry can be maintained in its original position or undergo a 180° degrees rotation around the x axis, depending on each case, so that in all cases studied the opening of the leakage always occurs at the bottom of the pipe. The case studies are listed in Table 1. The simulations were investigated using computers with Quad-Core Intel Xeon Processor E5430 Dual 2.66GHz with 8GB of RAM.

The leakage in 5 m from the fluid inlet section (hole diameter = 1.6 cm) is called Leak 1, the leakage in 7.5 m from the fluid inlet section (hole diameter = 1.6 cm) is called Leak 2 and the leakage in 5 m from the fluid inlet section (hole diameter = 3.0 cm) is called Leak 3.

RESULTS AND DISCUSSION

Fluid Volumetric Flow Rate Analysis

Figure 3 shows the evolution of the oil mass flow rate on the leak located at 5 m from inlet, with diameter of 1.6 cm, located in bottom side of pipe (Figure 1 rotated 180° around x axis). Figure 3(a) shows the evolution for cases where only the leak located at 5 m from inlet is activated (Cases 1, 3, 5, 7, 9 and 11) and the Figure 3(b) shows the same trend for the cases where the leaks at 5 m and at 7.5 m from inlet, with the same dimensions, are enabled (Cases 2, 4, 6, 8, 10 and 12). It is observed a similar behavior of the curves. With the decrease in volume fraction of oil entering the duct, it is observed a reduction of oil flow through the orifice leak. At the initial instants of the process, oil mass flow rate increase quickly until 0.4 s, when this time reaches the equilibrium condition (steadystate flow).

Figure 4 was done for continue the analysis of the oil mass flow rate at the hole cited before. It's notable that the presence of the second leak disturbs the behavior of the flow, changing the oil mass flow rate on the first leak. The solid line corresponds to the oil mass flow rate for the cases when only the leak at 5 meters from the input section is activated.

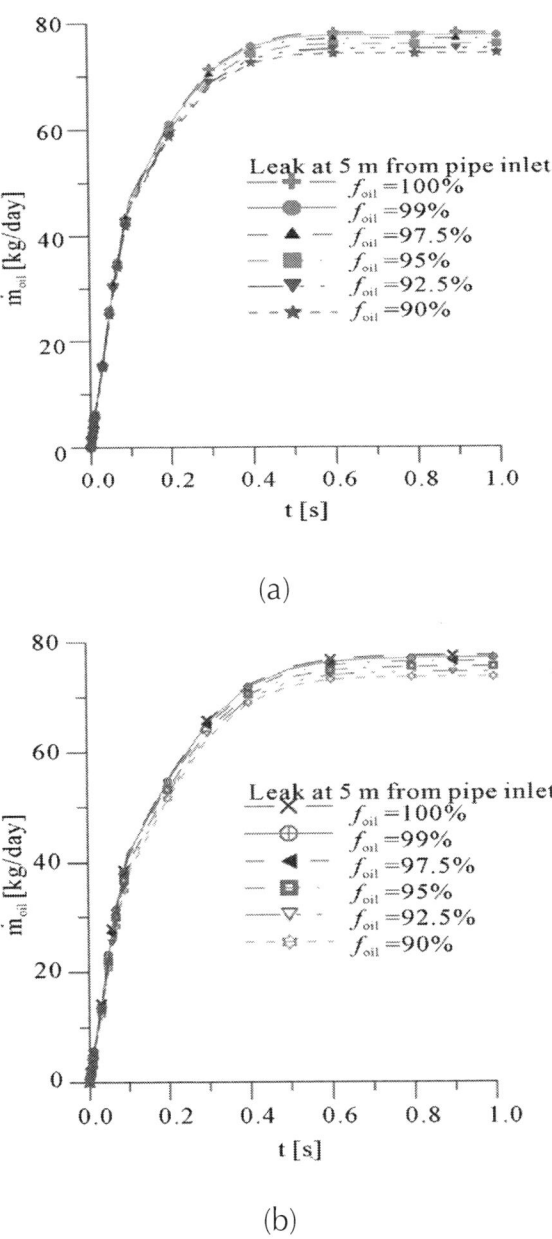

Figure 3: Oil mass flow in the orifice leak with diameter of 1.6 cm located at 5 m from the inlet versus time for different oil volumetric fractions injected into the pipe (a) Cases 1, 3, 5, 7, 9 and 11 and (b) Cases 2, 4, 6, 8, 10 and 12.

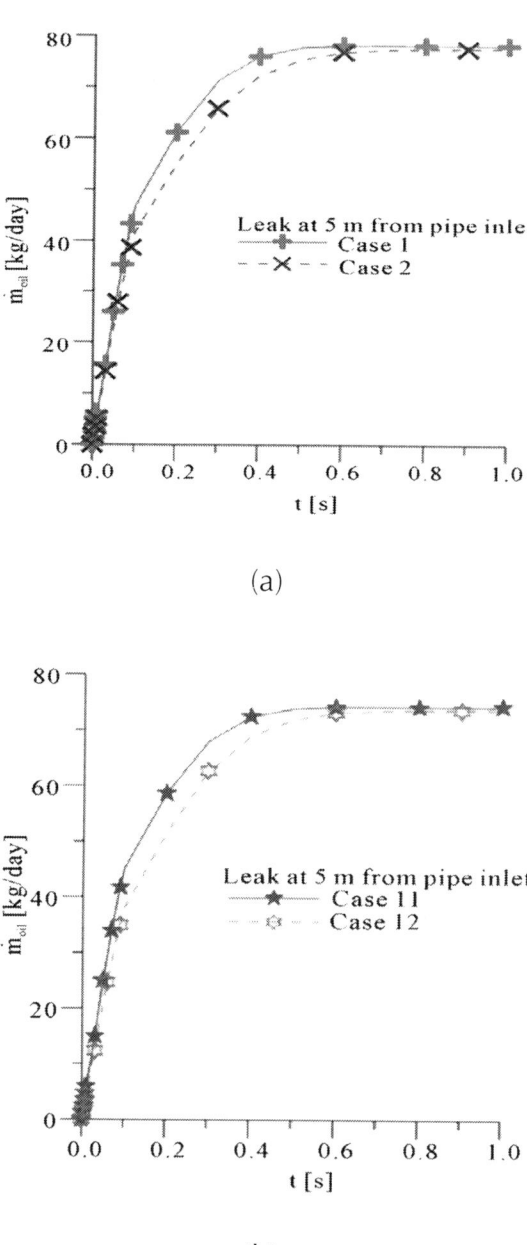

(a)

(b)

Figure 4: Oil mass flow rate in the orifice leak with diameter of 1.6 cm located at 5 m from the inlet versus time for different oil volumetric fractions injected into the pipe (a) Cases 1 and 2 and (b) Cases 11 and 12.

The dashed line corresponds to the oil mass flow rate in the same region, but when the first and second leaks are activated simultaneously. Comparing the two curves mentioned above, for the Cases 1 and 2 (Figure 4(a)), we see that the flow described by the continuous curve is stabilized higher and faster than the represented by the dashed curve flow. The same behavior is observed for the Cases 11 and 12 (Figure 4(b)).

For a better view of the graph (Figures 5 and 10), we adopted the dimensionless time as follows:

$$t^* = \frac{t}{t_{total}} \tag{5}$$

were t is the time and t_{total} is the total time of simulation. For the cases 01, 07 e 11 were adopted the total time of 1 second. For the cases 13, 14 and 15 the value is 1.8 s.

To analyse the effect of the orifice leak diameter on the oil mass flow rate, the cases 1, 7, 9, 13-15 are studied (Figure 5). To study the cases 13-15, where the leak is located in the bottom side of the pipe, the original position is maintained like the Figure 1 shows, i.e., the pipe is not rotated.

By increasing the diameter of the leak, it is observed that there is an increase in the oil mass flow rate in the leak, the continuous curve of Figure 5(a) is equivalent to the oil mass flow in Case 1, where the leak with a diameter of 1.6 cm and located at 5 meters from the input section is activated. The dashed curve refers to the oil mass flow rate in Case 13 where the leak with a diameter of 3.0 cm and located at 5 meters from the input section is activated. It is easily understood that the volumetric flow in the orifice with a diameter of 3.0 cm (dashed curve) is higher than for an orifice with a diameter of 1.6 cm (continuous curve). It is also interesting to note that by increasing the diameter of the orifice, more time is necessary to the oil mass flow rate to achieve the equilibrium condition. The same behavior is observed for Cases 7 and 14 (Figure 5(b)) and the Cases 11 and 15 (Figure 5(c)).

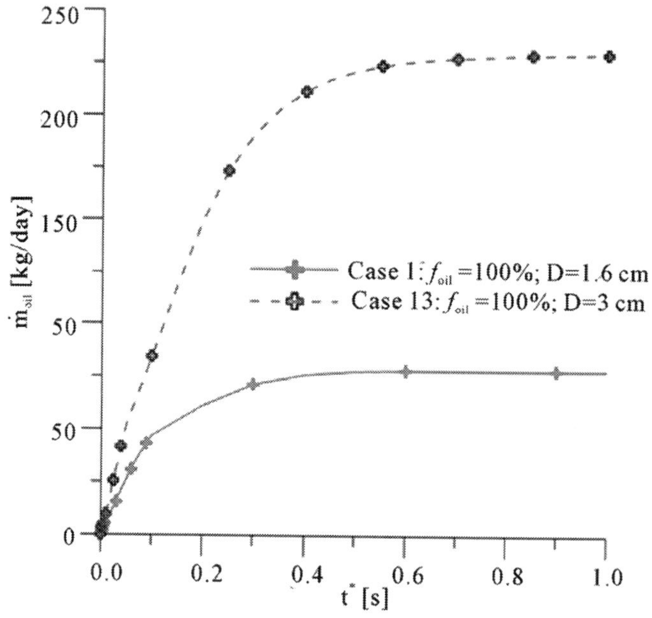

250
200
150
50
50
0
ṁoil [kg/day]
Case 1: foil =100%; D=1.6 cm
Case 13: foil =100%; D=3 cm
0.0
0.2
0.4
0.6
0.8
1.0
t* [s]

(a)

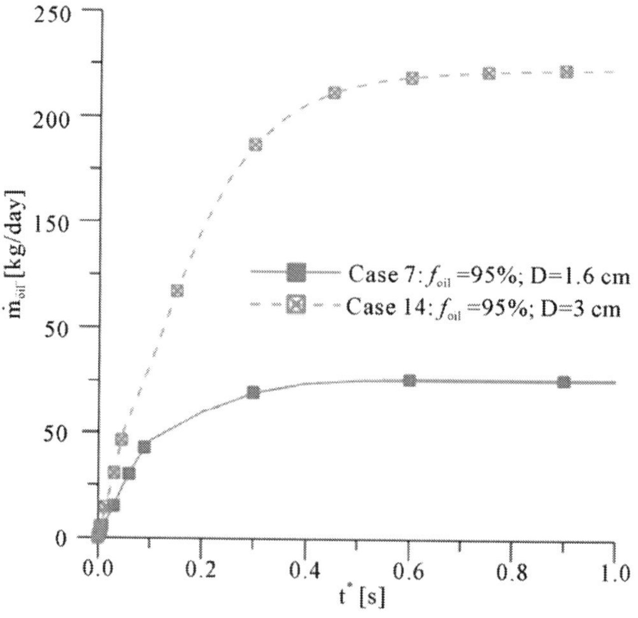

250
200
150
50
50
0
ṁoil [kg/day]
Case 7: foil =95%; D=1.6 cm
Case 14: foil =95%; D=3 cm
0.0
0.2
0.4
0.6
0.8
1.0
t* [s]

(b)

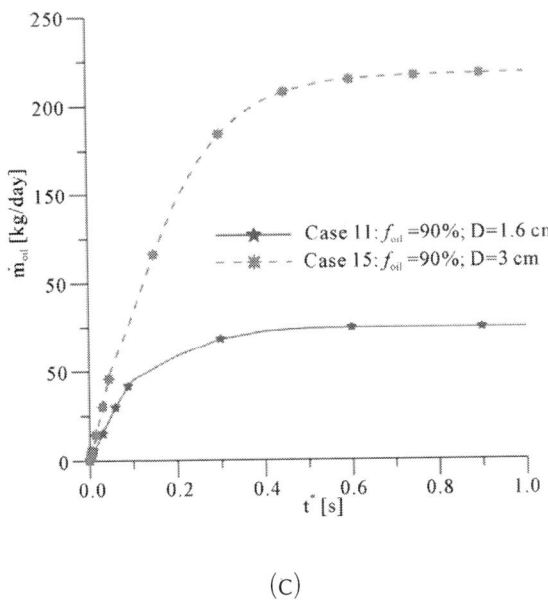

(c)

Figure 5: Oil mass flow rate in the leak versus dimensionless time for the leaks with different diameters.

Figure 6 illustrates the result of the total mass flow rate in leakage as a function of oil volumetric fraction at the entrance of pipe, after 1 second of flow for leak with a diameter of 1.6 cm and after 2 seconds of flow for leak with a diameter of 3.0 cm. It's possible to see that the smaller oil volumetric fraction in the mixture, the higher is the total mass flow rate leaving the orifice, which can be attributed to a reduction in viscosity of the mixture. As expected, the bigger the orifice leak the greater the loss of the fluids.

Pressure Drop Analysis

Figure 7(a) shows the evolution of the pressure drop between the entrance and the exit of the pipe as a function of time for different feeds of oil volume flow fractions when one leak is active (Cases 1, 3, 5, 7, 9 and 11). The similar behavior is observed, but with different magnitudes, for each value of the oil volumetric fraction and after 0.6 s of leaking there is a stabilization of the behavior of pressure. One fast drop in the pressure is verified at the initial instants of the flow due to presence of the leak.

The same behavior pressure drop between the inlet and outlet section is observed for cases where there are two active leaks, Figure 7(b) (Cases 2, 4, 6, 8, 10 and 12). However it is observed that the stabilization of the behavior of pressure occurs only after 0.7 s leak started. This behavior is explained by the fact two leaks in a same pipe cause a greater change as compared with the same pipe containing only one leak.

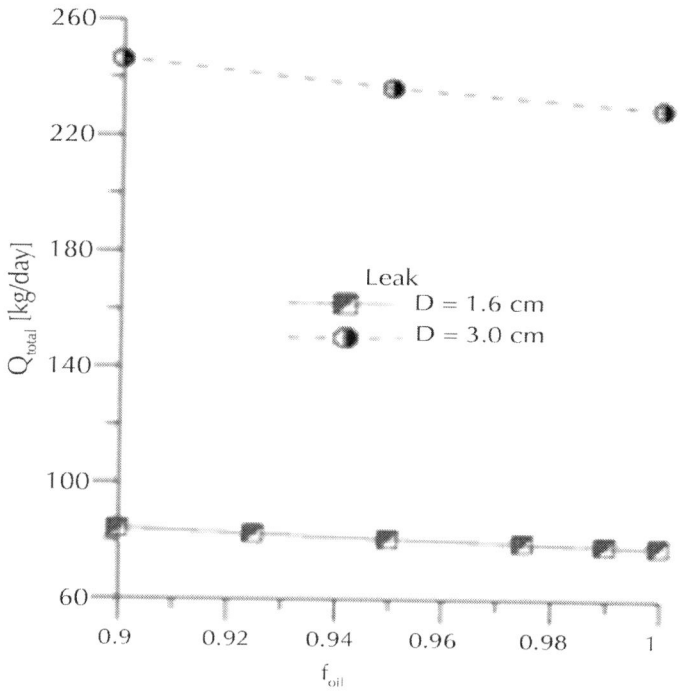

Figure 6: Oil-water mixture mass flow rate in the leak as a function of the oil volumetric fraction at the pipe inlet for leak with different diameters.

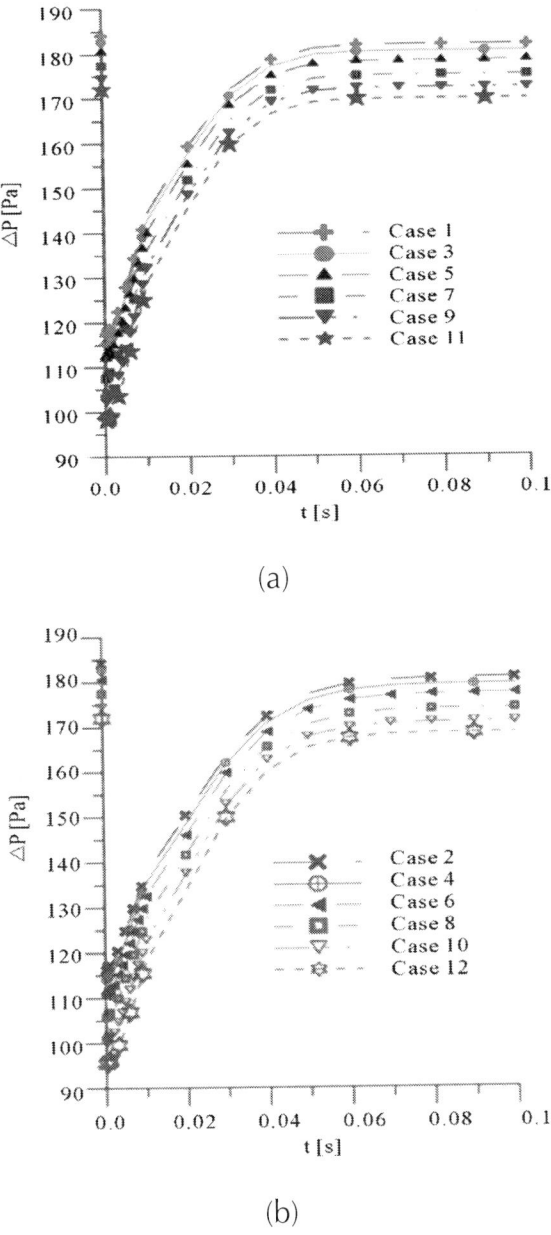

(a)

(b)

Figure 7: Pressure drop between the inlet and outlet sections of the pipe as a function of time and with presence of leakage (a) Cases 1, 3, 5, 7, 9 and 11 and (b) Cases 2, 4, 6, 8, 10 and 12.

In order to observe the behavior of pressure along the duct, taken up average pressure values in different planes yz distributed along the duct (0, 1, 2, 3, 4, 4.5, 5, 5.5, 6, 7, 7.5, 8, 9 and 10 m). With these average pressures were traced curves represented in Figure 8 for Cases 1, 3, 5, 7, 9 and 11 and Figure 9 for the cases 2, 4, 6, 8, 10 and 12 which shows the behavior of pressure along the duct. These curves were taken at the beginning of the leak (t = 0.001 s), t = 0.05 s, t = 0.1 s, t = 0.3 s and at the time that the behavior of the flow has been stabilized (t = 1 s). In all cases, the behavior change in pressure before and after the leak can be observed clearly.

By activating the leaks of 1.6 cm and 3 cm diameter, it is expected that they disturb the behavior of the pressure in the duct. The results have proven that after a certain period of time the pressure in the duct turn back to stabilize. Figure 10 (oil monophase flow) shows that by varying the diameter of the leak, the pressure stabilization period also varies.

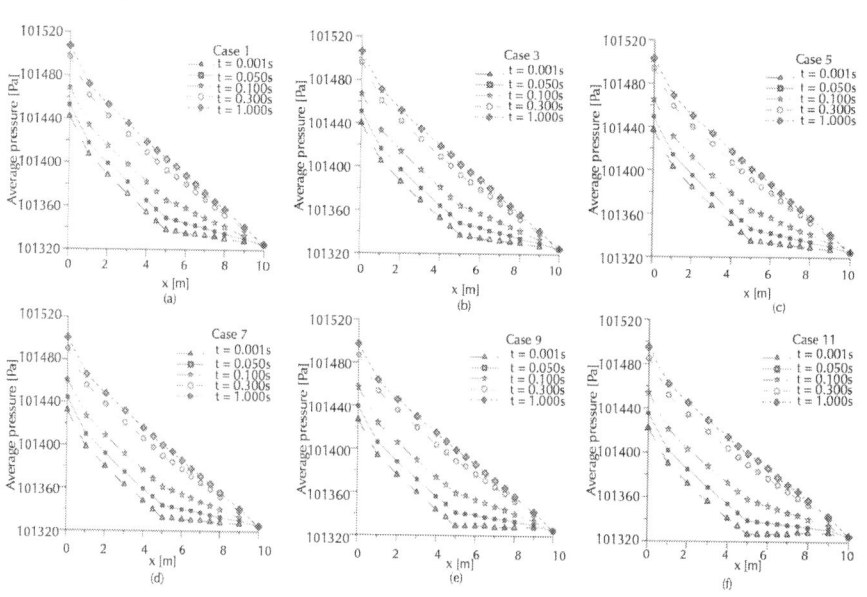

Figure 8: Average pressure along the duct at different flow time for the cases with only one leak.

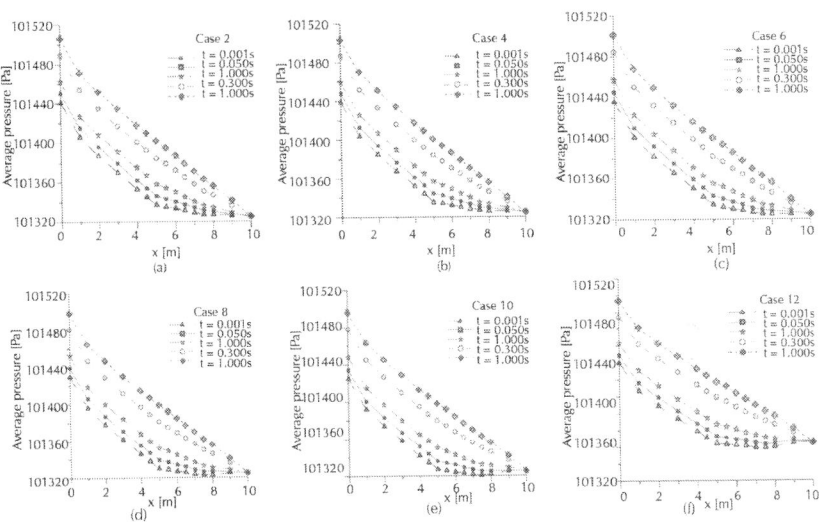

Figure 9: Average pressure along the duct at different flow time for the cases with two leaks.

The greater the leakage, the greater will be the disturbance, and therefore the time taken to stabilize the pressure in a new value.

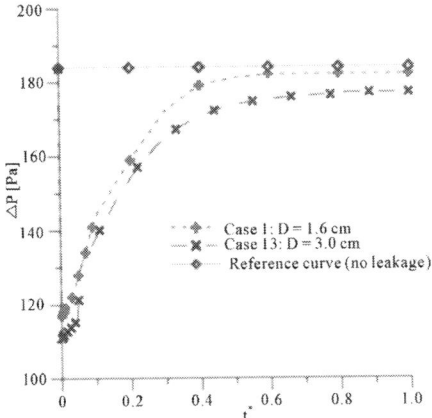

Figure 10: Average pressure drop between the inlet and outlet sections as a function of dimensionless time in the presence of leakage (Cases 1 and 13) and in the absence of leakage (reference curve).

CONCLUSIONS

With the results of numerical simulation of single-phase flow (oil) and two-phase flow (water-oil) in pipes, it can be concluded that:

- The proposed mathematical model was able to predict the leak in a horizontal pipe, evaluating the effect of the position, sizes and numbers of the leak on the fluid dynamic behavior of the mono and two-phase flows;
- The volume fraction of oil in two-phase mixture injected into the duct affected the amount of oil leaked;
- The presence of a second leak on downstream of the initial leakage on duct affected the oil flow of the first one, which became established with a lower flow rate compared to the situation where the first leak occured alone;
- By increasing the diameter of the leak orifice, the greater the volume of fluid leaving the orifice was;
- There was a slight variation of the pressure values before and after the leak and this varied depending on the fluid composition in the flow;

The time required for the pressure reaches stability after the leak depending on the size and locations of the leak in the pipe.

ACKNOWLEDGMENTS

The authors would like to express their thanks to Capes, CNPq, FINEP, PETROBRAS, ANP/UFCG/PRH-25 and RPCMOD for supporting this work, and also grateful to the authors of the references cited in this paper that helped in the improvement of quality.

REFERENCES

1.   J. Zhang, "Designing a Cost Effective and Reliable Pipeline Leak Detection System," Pipeline Reliability Conference, Houston, 19-22 November 1996, 11 Pages.

2.   C. M. Buiatti, "Monitoring of Tubes by Computational Thecniques On-Line," Dissertation, Department of Chemical Engineering, State University of Campinas, Campinas, 1995.

3.   F. M. de Azevedo, "Proposal of Algorithm for Leak Detection in Pipelines Using Frequencial Analysis of Signals Pressure," Dissertation, Federal University of Rio Grande do Norte, Natal, 2009.

4.   J. Agbakwuru, "Pipeline Potential Leak Detection Technologies: Assessment and Perspective in the Nigeria Niger Delta Region," Journal of Environmental Protection, Vol. 2, No. 8, 2011, pp. 1055-1061. http://dx.doi.org/10.4236/jep.2011.28121

5.   C. F. Braga, "Leak Detection by Computer 'On-Line' in Tubes Transporting Mixture Gás-Liquid," Master Dissertation, Department of Chemical Engineering, State University of Campinas, Campinas, 2001.

6.   B. A. F. Bezerra, "Leak Detection in Gas Pipelines by the Method of Transient Pressure Using CLP and Sensors," Monograph (Conclusion of Specialization Course), Federal University of Pernambuco, Recife, 2008.

7.   F. M. Garcia, J. Giaretton, M. B. Quadri and A. Bolzan, "Fluidodynamic Simulation of Leaking Water in a Section of Duct for Applications in Oil & Gas Industry," Proceedings of the 6th Brazilian Congress of Research and Development in Oil and Gas, Florianópolis, 9-13 October 2011, pp. 1-8.

8.   J. V. N. de Sousa, A. G. B. de Lima and S. R. de Farias Neto, "Numerical Analysis of Heavy Oil-Water Flow and Leak Detection in Vertical Pipeline," Advances in Chemical Engineering and Science, Vol. 3, No. 1, 2013, pp. 9-15.http://dx.doi.org/10.4236/aces.2013.31002

9.   A. L. Cunha, "Advanced Non-Isothermal Recovery of Heavy Oil in Petroleum Reservoirs by Numerical Simulation," Master Dissertation, Department of Chemical Engineering, Federal University of Campina Grande, Campina Grande, 2010.

10. J. R. Welty, C. E. Wicks and R. Wilson, "Fundamentals of Momentum, Heat, and Mass Transfer," 3th Edition, John Wiley & Sons, Hoboken, 1984.

Chapter 2

DomSign: A Top-Down Annotation Pipeline to Enlarge Enzyme Space in the Protein Universe

Tianmin Wang[1,3], Hiroshi Mori[1,2], Chong Zhang[3], Ken Kurokawa[1,2], Xin-Hui Xing[3], and Takuji Yamada[1]

[1]Department of Biological Information, Graduate School of Bioscience and Biotechnology, Tokyo Institute of Technology, 2-12-1 M6-3, Ookayama, Meguro-ku 152-8550, Tokyo, Japan

[2]Earth-Life Science Institute, Tokyo Institute of Technology, 2-12-1-E3-10 Ookayama, Meguro-ku 152-8550, Tokyo, Japan

[3]Department of Chemical Engineering, Tsinghua University, Beijing 100084, China

ABSTRACT

Background

Computational predictions of catalytic function are vital for in-depth understanding of enzymes. Because several novel approaches performing better than the common BLAST tool are rarely applied in research, we hypothesized that there is a large gap between the number of known annotated enzymes and the actual number in the protein universe, which significantly limits our ability to extract additional biologically relevant functional information from the available sequencing data. To reliably expand the enzyme space, we developed DomSign, a highly accurate domain signature–based enzyme functional prediction tool to assign Enzyme Commission (EC) digits.

Results

DomSign is a top-down prediction engine that yields results comparable, or superior, to those from many benchmark EC number prediction tools, including BLASTP, when a homolog with an identity >30% is not available in the database. Performance tests showed that DomSign is a highly reliable enzyme EC number annotation tool. After multiple tests, the accuracy is thought to be greater than 90%. Thus, DomSign can be applied to large scale datasets, with the goal of expanding the enzyme space with high fidelity. Using DomSign, we successfully increased the percentage of EC-tagged enzymes from 12% to 30% in UniProt-TrEMBL. In the Kyoto Encyclopedia of Genes and Genomes bacterial database, the percentage of EC-tagged enzymes for each bacterial genome could be increased from 26.0% to 33.2% on average. Metagenomic mining was also efficient, as exemplified by the application of DomSign to the Human Microbiome Project dataset, recovering nearly one million new EC-labeled enzymes.

Conclusions

Our results offer preliminarily confirmation of the existence of the hypothesized huge number of "hidden enzymes" in the protein universe, the identification of which could substantially further our understanding of the metabolisms of diverse organisms and also facilitate bioengineering by providing a richer enzyme resource. Furthermore, our results highlight the necessity of using more advanced computational tools than BLAST in protein database annotations to extract additional biologically relevant functional information from the available biological sequences.

BACKGROUND

Of the known biological sequences in the post-genomic era, the vast majority have not yet been, and cannot be, characterized by experimentation or manual annotation [1]. For example, Swiss-Prot, a protein database with a manually curated functional annotation, has only 547,085 entries as of December 2014, whereas a comprehensive protein database such as UniProt-TrEMBL, which contains a high-quality computationally analyzed functional annotations and covers most of the known protein sequences, contains tens of millions of members. Therefore, automated annotation is necessary to assign functions to uncharacterized sequences. Enzymes are of special importance owing to their central roles in metabolism and their potential uses in biotechnology [2]. Hence, a greater ability to predict enzyme functions will not only give biologists deeper insight into metabolism in general but also increase the toolkits for bioengineers.

Many novel bioinformatics tools with different bases, such as protein structure [3], functional clustering [4], evolutionary relationships [5] and biological systems networks [6], have been developed for enzyme or protein functional annotations. Many of them perform better [7], [8] than conventional approaches like BLAST, which is based on pairwise comparisons of gene sequence similarities to assign functions to new genes [9]. However, BLAST is currently the main approach used in functional annotations [10], whereas many recently developed tools are rarely applied in research projects [7]. Additionally, BLAST-based functional annotations perform poorly when only distantly related homologs with similarities of <30% can be found [11], [12]. Furthermore, many proteins recently discovered using metagenomics approaches do not have homologs with high enough amino acid sequence identity levels for reliable functional annotation. For example, in a benchmark study, which used a metagenomics approach focusing on cow rumen–derived biomass-degrading enzymes, it was found that in terms of amino acid sequence identity, only 12% of the 27,755 carbohydrate-activated genes assembled had >75% identity to genes deposited in NCBI-nr, whereas 43% of the genes had <50% identity to any known protein in NCBI-nr, NCBI-env and CAZy [13]. Thus, if novel and combinatorial approaches are used, to what extent, with acceptable precision, can we improve the coverage of the protein annotation? For enzymes, there is a well-established system, the Enzyme Commission (EC) number [14], which describes catalytic functions hierarchically using four digits. As far as we know, although many EC number prediction tools are available, most are limited to performance tests within small datasets and none of them has been used to systematically address the comprehensiveness of enzyme functional annotation in public protein database. Thus, a more specific question, "To what extent we can improve, with an acceptable precision, the coverage of enzyme annotations using EC numbers?" is worth addressing by illustrating the power of approaches whose utility goes beyond BLAST. The insight we obtain can be also generalized to protein annotations for other functional attributes.

Thus, novel approaches with high coverage rates that maintain an acceptable precision are of special interest. Hierarchical or top-down algorithms with a layer-by-layer logic satisfy these requirements [15], [16]. Such approaches assign functions only at a level that can be inferred with high confidence. Hence, in many cases, general rather than specific functions (for example, the top level of EC numbers) are assigned to avoid the overprediction of protein functions, such as annotation below the trusted cutoff or inference only from a superfamily, a main problem of current database annotations [17]. Furthermore, this approach is suitable for widely accepted protein function definition systems, such as EC or Gene Ontology (GO), both of which are widely applied metrics systems to consistently describe the functions of gene products [18], owing to their hierarchical structure.

Domains are conserved parts of a given protein's amino acid sequence and structure that can evolve, function and exist independently of the rest of the amino acid chain. Thus, it has been hypothesized that machine learning with domains as input labels might serve as a powerful approach to predict protein functions [19]. For example, the dcGO database, based on associating SCOP domains or domain combinations with GO terms of protein products, infers the domain or domain combinations responsible for particular GO terms [20]. A domain architecture–based approach might thus be a powerful tool for predicting enzymatic functions. Here, we report on "DomSign", a top-down enzyme function (EC number) annotation pipeline based on domain signature–derived machine learning. We must emphasize, based on the belief that any reliable protein function prediction tools should depend on multiplicity [21], that our purpose here is not just to present a simple function prediction tool but rather to address the issue of to what extent can the coverage of enzyme annotations by EC numbers be improved, with acceptable accuracy, by methods beyond simple BLAST.

To test the reliability of DomSign, many benchmark enzyme annotation methods were compared with. The performance of DomSign was comparable, or superior, to all of them after exhaustive testing against reliable datasets, such as Swiss-Prot enzymes, suggesting that DomSign is a highly reliable enzyme annotation tool that can identify more enzymes in the protein universe. Furthermore, to expand the number of enzymes retrieved from large datasets, we compared our results with those proteins already assigned EC numbers in the original dataset. More 'hidden enzymes' were predicted by DomSign. Thus, DomSign, with

>90% accuracy suggested by the tests, can be used to predict a large number of enzymes by assigning EC numbers to proteins in both the UniProt-TrEMBL [22] and Kyoto Encyclopedia of Genes and Genomes (KEGG) [23] bacterial subsection, which, respectively, represent the most complete protein database and best metabolic pathway information collection. DomSign also can be applied to metagenomic samples as exemplified by the Human Microbiome Project (HMP) dataset [24], a comprehensive and well-analyzed metagenomic gene dataset focused on parsing the interactions between commensal microorganisms of humans (human microbiome) and human health. In this case, DomSign not only significantly increased the number of EC-labeled enzymes but also helped to clarify the metabolic capacity of the sample by recovering new EC numbers beyond the official annotation. These results highlight the necessity to develop enzyme EC number prediction projects or, more generally, protein annotation projects with novel approaches akin to DomSign to extract more biological information from the available sequencing data.

METHODS

Definition of a Domain Signature

Pfam is a protein domain collection with ~80% coverage of the current protein universe [25], and its Pfam-A subsection is highly reliable owing to its manually curated seed alignment. For our purpose, a string of non-duplicated Pfam-A domains belonging to a protein was defined as its domain signature (DS) and used to predict function(s). Although some research has suggested a potential advantage of involving domain recurrence and order in protein GO assignments [26], our results showed that this simpler DS definition provided a higher coverage for proteins identified in metagenomics studies. When utilizing Swiss-Prot protein DSs to retrieve HMP phase I non-redundant proteins, the coverage was 74.7% when considering domain recurrence and order versus 77.1% with more simple definition. Unlike the GO term assignment used previously [26], recurrence did not lead to a significant difference in coverage as indicated by reconstructing the EC number machine-learning prediction model (1) used in this work, whose method is presented in the following part. Thus, because the main aim of our study was to improve enzyme annotation coverage, our simpler DS definition was applied.

Preparation of the Dataset

Swiss-Prot and TrEMBL datasets were downloaded on November 2, 2013, from the Pfam ftp site (version 27.0) from which Pfam-A domains were extracted. Pfam-A Hidden Markov Model (PfamA. hmm) for hmmsearch (version 3.1b1) [27] was accessed from the same site. The HMP phase I non-redundant protein dataset (95% identity cut-off, 15,006,602 entries from 690 samples) [24] was collected from the HMP data processing center (http:// www.hmpdacc.org/). A benchmark dataset for unbiased tests was collected from [15] (Supplementary Data 2). The files (gene IDs and sequences in the fasta format) from KEGG were downloaded on March 6, 2014. The EC2GO mapping file [28] was downloaded on June 20, 2014 from the GO homepage. All of these files were further processed as stated below.

"Sprot Enzyme" Dataset

The Swiss-Prot dataset is a protein collection with an exhaustive manually curated—and thus reliable—functional annotation. In this context, it was a good choice working as the training set for comparing prediction model performance by cross-validation. The subset of enzymes in Swiss-Prot with both single EC numbers and Pfam-A domains was termed "sprot enzyme", encompassing 228,710 entries and 4,216 distinct DSs. This set was used to construct the "Specific enzyme domain signature" dataset as described below and also as a training dataset to build the prediction model for enzyme mining in several general protein databases (TrEMBL, KEGG and HMP).

"Sprot Protein" Dataset

Another subset of Swiss-Prot, which contains all of the Pfam-A proteins with single or no EC numbers, was named "sprot protein", encompassing 46.8% enzymes (with single EC numbers) and 53.2% non-enzymes (without EC numbers), which covers 99.0% of the Swiss-Prot proteins with Pfam-A domains. This dataset was used for model parameter optimization and performance comparisons against BLAST and FS models (see descriptions below in Methods about FS model) [19].

"Specific Enzyme Domain Signature" Dataset

To identify enzymes from the protein pool, we further constructed a "Specific enzyme domain signature" dataset. The fundamental idea was to remove non-enzyme-derived DSs from the 4,216 distinct DSs belonging to "sprot enzyme". Because EC numbers do not cover all enzymes, however, a more reliable non-enzymatic dataset beyond simple proteins without EC numbers needed to be constructed. Briefly, for the proteins without EC numbers in Swiss-Prot, their annotation raw files ('KW', 'DR' and 'DE' lines) were filtered using a catalytic or functional uncertainty–inferring term ('iron sulfur', 'uncharacterized', 'biosynthesis', 'ferredoxin', 'ase', 'enzyme', 'hypothetic', 'putative' and 'predicted') to reliably extract non-enzymes. By this means, we collected 2,901 unique DSs from 157,240 non-enzymes carrying Pfam-A domains. After removing these DSs from the "sprot enzyme" DS set, 3,949 specific enzyme DSs were acquired, covering 95.4% of "sprot enzyme". This dataset was used for selecting enzyme candidates from a protein pool using the benchmark comparison method and enzyme mining process.

"SVMHL Unbiased" Dataset

To compare the performance of our approach with the SVMHL pipeline (see descriptions below in Methods about SVMHL model) [29], the aforementioned unbiased dataset was further processed to remove, as described in their paper, enzyme sub-subfamilies with fewer than 50 members.

"TrEMBL Enzyme" and "HMP Enzyme" Datasets

The TrEMBL raw dataset was filtered to extract enzymes with single EC numbers and Pfam-A domains, producing "TrEMBL enzyme". Likewise, "HMP enzyme" was constructed from the HMP non-redundant protein set. Pfam-A domains were retrieved by an hmmsearch against PfamA.hmm using the cut_tc cutoff with all other parameters set as default. These two datasets were used as the gold standards to test the reliability of the DomSign-based enzyme EC number annotation prior to the actual enzyme mining of TrEMBL and HMP original datasets. The statistics and usage of the datasets constructed in this work are presented in 2.

Prediction Model Description

Our prediction model consists of two separate steps: enzyme differentiation from the protein pool and EC number annotation based on machine learning. In the first step, proteins in query datasets are recognized as potential enzyme candidates if their DSs are among the aforementioned "Specific enzyme domain signature" set. In the second step, a top-down machine-learning model is developed to predict EC numbers.

First, we converted the training dataset into a list in which every protein had one DS and one EC number (Figure 1(1)). Subsequently, the proteins were categorized based on their DSs.

Thus, we constructed a series of protein groups in which all members contained the same DS. Here, we define the number of member proteins in one group as N_{DSi}. Then, the member proteins in one group were further divided into subgroups based on their EC numbers, leading to a protein subgroup with the same EC (N_{DSi} _ ECj and $N_{DSi} = \sum_j N_{DSi}\text{-}ECj$) (Figure 1(2)). Further, the abundance of every subgroup among one protein group was calculated ($A_{DSi\text{-}ECj} = N_{DSi\text{-}ECj}/N_{DSi}$) (Figure 1(3)). In each group, there exists at least one dominant subgroup with the highest abundance. The EC number for this subgroup is then associated with the relevant DS, whereas the abundance of this subgroup is defined as the "specificity" for this DS-EC pair, which acts as the fundamental parameter in the machine-learning model (Figure 1(4)). We constructed four prediction models to assign four levels for one complete EC hierarchy. For each model, at the first step (Figure 1(1)) one fraction of the EC number was extracted—for instance, for the model focusing on the second EC hierarchy, EC = x.x.-.- is extracted. All further steps were the same during the construction of these four models. Thus, this machine-learning approach makes it possible to annotate the EC hierarchy from general to specific where the "specificity" of DS-EC pairs can be used to balance the tradeoff between recall and precision, depending on the particular purpose.

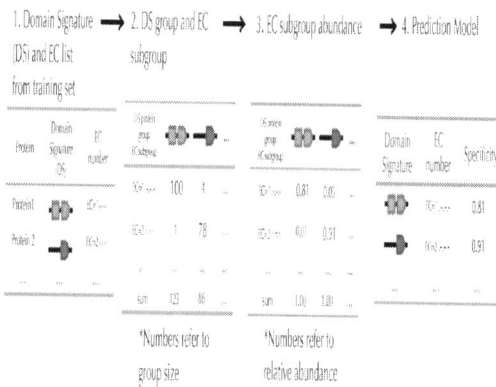

Figure 1: Construction of the machine-learning model to predict EC numbers. (1) Test dataset: DSs and EC numbers for every enzyme were ex-

tracted from original datasets, such as Swiss-Prot. (2) These proteins were categorized into groups based on common DSs. Subsequently, the groups were divided into subgroups based on the corresponding EC numbers. Thus, the numbers in each cell represent the number of proteins in each subgroup, and the total member number for each group is summarized in the last row. The numbers of dominant subgroups within one group are colored red. (3) The abundance of each subgroup within its parent group (the same DS) was calculated and represented. The abundance of dominant subgroups for each group (the same DS) is colored red. (4) Prediction model: Every DS was associated with the relevant dominant EC number within its protein group (carrying this DS). The abundance of dominant EC subgroups was extracted and set as the "specificity" for this EC-DS pair.

From Model to Prediction Engine

First, the training dataset was used to construct four prediction models for each EC hierarchy level, and the DSs of query proteins were calculated by hmmsearch with a cut_tc cutoff and all other parameters set as default. Then, the specific enzyme DS dataset was used to select potential enzyme candidates from query proteins. Then, four constructed prediction models were used one by one to annotate EC digits, assigning the EC number that corresponds to the query DS. In this process, a specificity threshold is applied to balance precision and recall. Specifically, when the "specificity" of the DS-EC pair is less than the specificity threshold, the procedure is shut down and only the EC digits annotated previously form the output (Figure 2). In this way, the precision can be increased by making the specificity threshold stricter with a loss of recall, or vice versa. Additionally, although it is not statistically rigorous, the specificity for one particular DS-EC pair can be used as the confidence score to infer the reliability of each prediction by DomSign. For example, if DomSign assigns one enzyme with EC number 1.1.1. - and the specificity values for the DS-EC pair of the first three hierarchical levels are 0.9, 0.88 and 0.85, we can simply set these three parameters as the confidence score for the reliability of prediction for the first three EC digits, respectively. The script package for this tool is provided as 3.

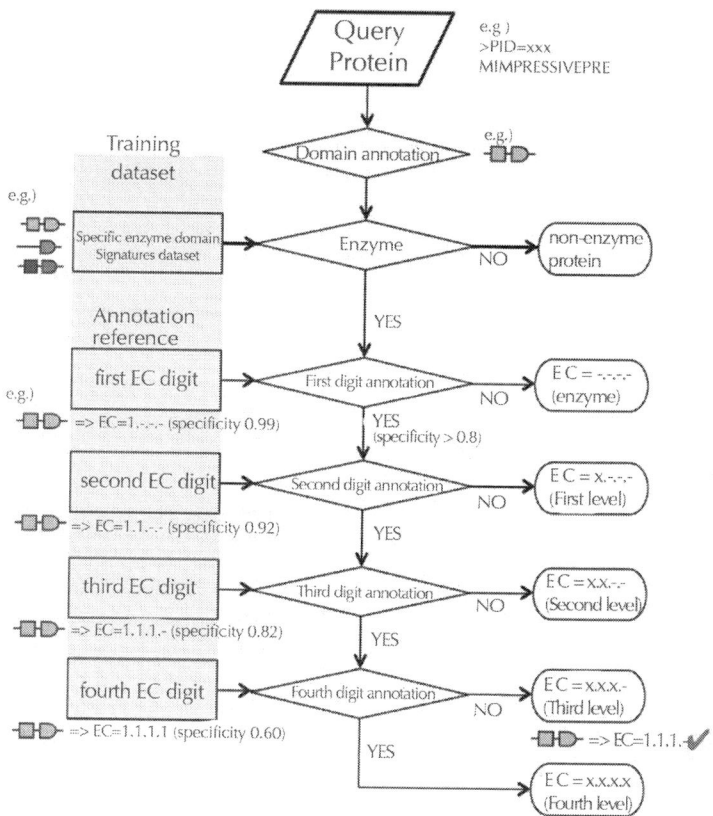

Figure 2: Schematic representation of the DomSign pipeline. The pipeline is divided into two parts—enzyme candidate selection and EC number annotation. In the first step, specific enzyme DSs are utilized, and all proteins with DSs within this dataset are selected as potential enzyme candidates. Simultaneously, four annotation references for the EC digits at four levels are constructed as described in Figure 1. At every level, if the "specificity" of the corresponding DS-EC pair in the annotation reference is less than the user-defined threshold, the pipeline is shut down and the previously annotated EC digits form the output. If not, the pipeline continues until the fourth EC digit has been annotated. An example of the DomSign procedure to annotate a protein is shown here. Because the specificity threshold is above the specificity of the DS-EC pair at the last level, only the first three DS-EC digits are predicted, leading to final result: EC = 1.1.1.-.

Performance Evaluation Statistics

Owing to the top-down nature of our approach, we designed a new result evaluation system to use instead of the widely used recall-precision curve [19] that differentiates the annotation results at different levels, resulting in better resolution. Briefly, the predicted EC number (PE) is compared with the true EC number (TE), and the result is classified into the following groups (Figure 3A, right): E—Equality, PE is the same as TE ("PE: EC = 1.2.3.4" vs. "TE: EC = 1.2.3.4"); OP—Over prediction, there is at least one incorrectly assigned EC digit in PE compared with TE ("PE: EC = 1.2.1.1" vs. "TE: EC = 1.2.3.4"); IA—Insufficient Annotation, PE is correct but not complete compared with TE ("PE: EC = 1.2.-.-" vs. "TE: EC = 1.2.3.4"); and IM—Improvement, TE is the parent family of PE ("PE: EC = 1.2.3.4" vs. "TE: EC = 1.2.3.-"). When TE is "Non-enzyme", if the PE equals "Non-enzyme", then the comparison result is set as "Equality". Otherwise, the result is "Over prediction". Additionally, if PE is "Non-enzyme" and TE is not, then the comparison result is set as IA. What needs to be specifically mentioned here is IA. Although this result means incomplete annotation, it is correct and does not cause any increase in the error rate. Thus, IA provides better annotation coverage and simultaneously maintains high precision. The evaluation metrics defined here differ from traditional ones [19]. However, compared with previous precision-recall curves that equally consider different EC hierarchy levels, this system covers all the possible situations and also gives an intuitive view of the performance at different annotation levels with higher resolution, which is especially suitable for evaluating annotation results using metrics of a hierarchical structure.

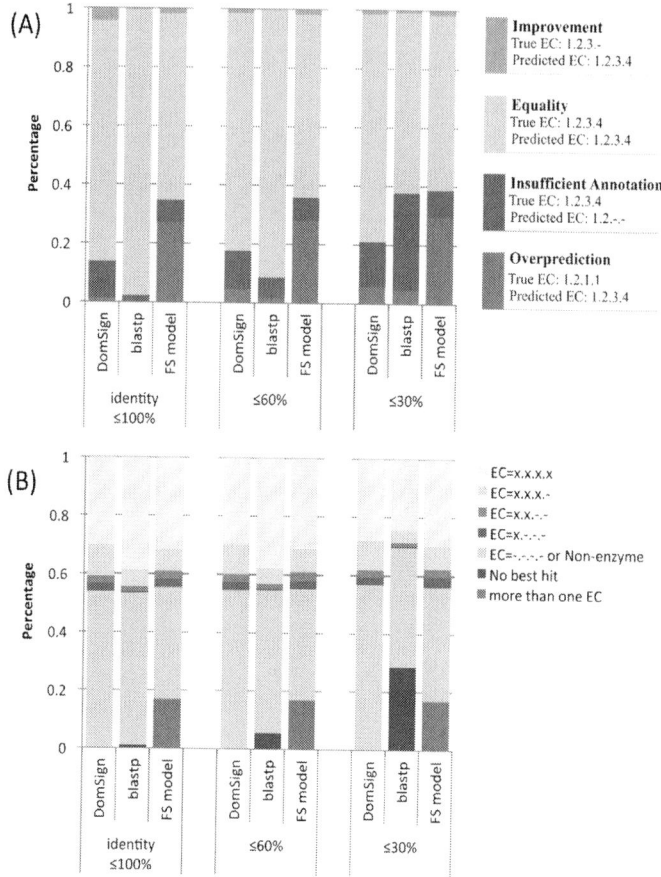

Figure 3: DomSign performance comparison with BLAST and FS models by 1,000-fold cross-validation of "sprot protein". Three levels of 1,000-fold cross-validations were conducted for each method. Homologs of a query above a given threshold ("identity $\leq$ 100%", "identity $\leq$ 60%" and "identity $\leq$ 30%" described in Methods) were removed from the reference dataset and, for each reference dataset, only sequences below the given threshold were kept. In this test, an 80% specificity threshold, 10^{-3} E-value and default parameters were applied to the DomSign, BLASTP and FS models. The relative standard errors were not significant (<1%) and therefore are not illustrated here. (A) Results for the evaluation of the three methods. As shown on the right, four attributes are defined to evaluate the annotation results in contrast to the "true EC number" (see Methods for details). (B) The EC hierarchy level distribution in the annotation

results of the three methods. Seven attributes are defined here to describe the annotation results. Among them, "No best hit" is specific to BLASTP. "More than one EC" is specific to the FS model because this dataset encompasses only enzymes with single EC numbers or non-enzymes, and this attribute is regarded as "OP" in Panel A. We integrated the annotation result "Non-enzyme" and "EC = −.-.-.-", as shown in Figure 2 into one unified group, "Non-enzyme", in the result's illustration because the latter has no EC number assigned and also only occupies a small fraction of the annotation results (the ratio of the "EC = −.-.-.-" subclass is only 1.4% in the "identity ≥ 100%" group for DomSign) of the annotation results.

Performance Test

Four benchmark methods, BLASTP (2.2.28+), FS [19], SVMHL [29] and EnzML [30], were selected to test the performance of DomSign.

Comparison with BLASTP and FS by Cross-validation

For the FS model, the script package from Forslund K. et al. [19] was run on our system to calculate the GO terms derived from the DS defined in their work. Subsequently, we used the EC2GO mapping file to convert the FS model's predicted GO terms to EC numbers. If multiple EC numbers existed for one particular GO term, we assigned that protein all of the relevant associated EC numbers. The three pipelines were tested by 1,000-fold cross-validations of the "sprot protein" dataset. Because the dataset has only enzymes with single EC numbers or non-enzymes, if the FS model predicted more than one EC number for a query then the result was "OP". Furthermore, to simulate the situation in which no sequences in the database have a high similarity to the query protein, two additional rounds of cross-validations against "sprot protein" were executed. Briefly, sequences in the training set having specificities above threshold I (60% identity, 80% query coverage) and II (30% identity, 80% query coverage) with any query sequence, respectively, were removed by BLASTP. In this way, any sequence in the training set is no more similar to any query sequence than the defined threshold. These two rounds of cross-validation, together with the common cross-validation, were termed "identity ≤ 100%", "identity ≤ 60%" and "identity ≤ 30%". For BLASTP, a 10^{-3} E-value and default parameters were applied. For the FS model, parameters were set as default for the processing.

Comparison with the SVMHL Model by Cross-validation

Because the source code of SVMHL is not available, we compared DomSign with SVMHL by the same test as stated in [29], and the raw data were used for performance comparisons. Briefly, a 10-fold cross-validation was conducted using DomSign on the "SVMHL unbiased dataset", and prediction accuracy [29] was used to evaluate the results. In this case, accuracy is defined as the percentage of completely correct annotations. Here, one predicted EC number at one specific hierarchy level (an EC number consisting of three digits when the EC hierarchy level is three) is set as 'correct' when its component digits are all correct. Because SVMHL does not have an enzyme and non-enzyme differentiation step, we included only the predicted "enzyme" by DomSign in the results comparison, which covered 85.2% ± 0.4% of the query proteins on average during the cross-validation.

Comparison with EnzML

Like the SVMHL model, owing to the inability to run EnzML on our system, we also compared the performance between EnzML and DomSign by the same test stated in [30], and the data published in that paper were used as the benchmark. The "Swiss-Prot&KEGG" set and the less redundant "UniRef50 Swiss-Prot&KEGG" set were constructed according to the description in the EnzML paper [30], and a 10-fold cross-validation was conducted. The example-based precision and recall rate were applied to the performance evaluation. Briefly, these two metrics consider how many correct EC predictions are assigned to each individual protein example on average [31]. For example, for each protein, true (TE) and predicted (PE) EC number sets at every hierarchical level (EC = 1.1.- . - is decomposed to EC = 1, EC = 1.1, EC = 1.1. - and EC = 1.1.-.-) are extracted and compared with each other. The example-based precision and recall rate can be defined by the two equations shown below:

$$Precision = \frac{1}{m} \sum_{i=1}^{m} \frac{|TE_i \cap PE_i|}{PE_i}$$

$$Recall = \frac{1}{m} \sum_{i=1}^{m} \frac{|TE_i \cap PE_i|}{TE_i},$$

Here 'm' refers to total number of proteins, and TE_i and PE_i refer to the sets of annotated EC numbers at four hierarchical levels or 'Non-enzymes' for each protein.

Enzyme Predictions from Large-scale Datasets

"Sprot enzyme" was used as the test dataset, and "Specific enzyme domain signature" was used to select enzyme candidates. "TrEMBL enzyme" and "HMP enzyme", combined with their original annotations, were used to evaluate the reliability of DomSign for expanding enzyme space. All TrEMBL and HMP proteins were then annotated by DomSign to test the extent of the enzyme expansion. Further, to show the significance of enzyme expansion in KEGG, among the predicted novel enzymes of TrEMBL, novel enzymes for 2,584 bacterial genomes in KEGG were extracted. Owing to the subtle differences between KEGG and TrEMBL annotations, a few novel enzymes in TrEMBL have EC numbers in KEGG. These were removed to retrieve the exact number of novel enzymes from KEGG, and the relevant statistics were calculated.

RESULTS

Optimization of the DomSign Specificity Threshold

We tested the reliability of DomSign as an EC number prediction tool. Because we designed a parameter "specificity threshold" (Methods) in DomSign to balance the tradeoff between precision and recall (Figure 2), three rounds of 1,000-fold cross-validations ("identity≤100%", "identity≤60%" and "identity≤30%" cutoffs as described in Methods) were performed on the "sprot protein" dataset using DomSign with 99%, 90%, 80% and 70% specificity thresholds to optimize this parameter (4). Among the 99%, 90% and 80% specificity thresholds, the 80% had the best coverage (IA, E and IM) and a slightly increased error rate (OP). However, further reduction of the specificity threshold to 70% resulted in a much smaller increase in coverage accompanied with a relatively severe OP ratio, especially for the "identity≤30%" group, indicating that 80% might be the optimal specificity threshold for DomSign. Thus, we applied this parameter in further analyses.

Comparisons among DomSign, BLAST and FS Models

BLAST was selected as the benchmark because of its wide application in research, and we used the best hit of BLAST to assign EC numbers. The FS model applies similar DS definitions, with no consideration for recurrence or order. However, it considers the contributions of every subset of DSs rather than regarding them as intact labels. Briefly, this model utilizes Bayesian statistical methods to evaluate the possibility of one particular GO annotation term inferred from all the subsets of the DS. By averaging the contributions of all the subsets, the probability of one protein having this annotation term can be calculated accordingly. There are three reasons for the comparison with the FS model: first, it utilizes domain information to assign GO terms. Thus, it can act as a good benchmark among the domain architecture–based methods. Secondly, this method yields reliable GO assignments, even in the situation where UniRef50 is applied for cross-validation, indicating the performance stability in an unbiased condition; and finally, the FS model provides a very user-friendly package for command line usage. Here, we converted GO terms to EC numbers using the EC2GO mapping file provided by the GO consortium [28].

Similar to the last section, to compare the performance among DomSign, BLAST and FS models, especially when the database contained no sequences having high similarities to the query protein, three rounds of 1,000-fold cross-validations ("identity ≤ 100%", "identity ≤ 60%" and "identity ≤ 30%" as described in Methods) were conducted on the "sprot protein" dataset by DomSign with an 80% specificity threshold, BLASTP with a 10^{-3} E-value and the FS model with default parameter settings. It is necessary to emphasize the importance of performance tests using this scenario because BLAST itself performs enzyme functional annotations well (above 90% precision and recall in some situations) when homologs with similarities above a particular threshold are available [12]. Thus, there is limited room for further improvement in this regard, whereas there is ample need for improvement when homologs are unavailable.

With the accumulation of novel sequences, this issue is expected to become more important. Thus, in the development of a new generation of computational approaches, more attention should be paid to the "homolog unavailable scenario". As shown in Figure 3, machine learning–based methods, such as DomSign and the FS model, are much more robust when there is a reduced homolog availability compared with BLAST. Meanwhile, with a significant increase in "No best hit" (Figure 3B), coverage for BLAST decreases dramatically. Hence, in contrast to the nearly perfect performance of BLAST in the "identity ≤ 100%" group, DomSign achieved an overall performance superior to BLAST in the case of "identity ≤ 30%", producing a comparable OP ratio but much higher coverage. Meanwhile, the FS model tended to have a very high OP ratio in all three tests, partly because of the multiple EC number predictions (Figure 3A) in this single EC enzyme plus non-enzyme dataset (2) and partly because of incorrect EC assignments (both reasons contributed ~50% to the high OP level in the FS model, Figure 3A, B). Therefore, DomSign has the potential to partly replace BLAST as a functional annotation tool for novel proteins that have no homologs in the database.

Comparison with SVMHL using an Unbiased Dataset

To further test the effectiveness of DomSign with respect to avoiding potential bias towards abundant enzyme families [32], the "SVMHL unbiased dataset" was subjected to a 10-fold cross-validation because any two sequences have <50% identity and the enzymes are manually selected to cover most of the enzyme families without bias. The SVMHL model [29] is the benchmark that annotates EC hierarchy by considering two main features, namely the abundance of every possible tripeptide sequence within a polypeptide [33] and a protein structure–based enzymatic function prediction model. The annotation accuracy of DomSign and SVMHL at the second and third EC hierarchy levels is shown in 5. Although the accuracy for the SVMHL model at the second hierarchy level was slightly greater than that of DomSign, at the third hierarchy level DomSign outperformed SVMHL for most enzyme families. Because Wang et al. [29] did not present their results at the fourth level, only the DomSign results at this level are shown (5). Based on this comparison, DomSign works well in the unbiased situation compared with other benchmark methods.

Comparison with EnzML

The EnzML model is a multi-label classification method that uses Binary Relevance Nearest Neighbors (BR-kNN) to predict EC numbers [30]. Briefly, this model utilized a more general protein signature set, InterPro [34], rather than Pfam as the input label. A multi-label support vector machine methodology was used, and the k parameter—the number of neighbors considered during the prediction—was optimized to '1'. The methodology of the multi-label support vector machines can be intuitively considered as the combination of multiple support vector machines for a series of binary labels ('yes' or 'no' for one particular EC hierarchy). Noteworthy, Mulan [35], an open-source software infrastructure for evaluation and prediction, is used for this specific work. This model is presently the best benchmark, which has been shown to be superior to some other widely used tools such as ModEnzA [36] and EFICAz2 [37]. "Swiss-Prot&KEGG" and the less redundant "UniRef50 Swiss-Prot&KEGG" [30] datasets were used for the 10-fold cross-validation (Figure 4A, B). Although the differences were not significant, we observed that EnzML performed better than DomSign in terms of example-based precision and recall. To clarify the source of these differences, for our evaluation we excluded the real enzymes that were incorrectly predicted as non-enzymes by DomSign (Figure 4C, D). Thereafter, DomSign's performance became comparable to that of EnzML.

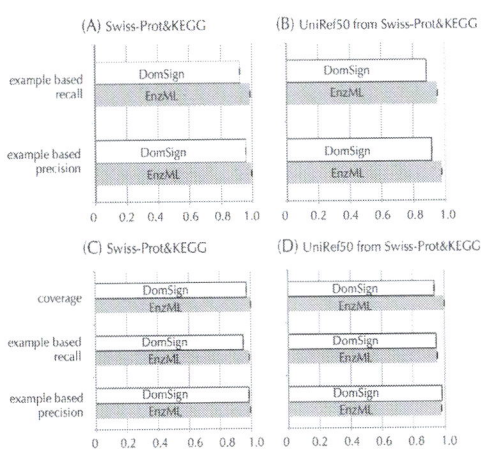

Figure 4: Comparison between DomSign and EnzML using Swiss-Prot&KEGG and Swiss-Prot&KEGG extracted by UniRef50 datasets. The

barplot represents accuracy caluclated by DomSign (white) and EnzML (gray). In contrast to panels (A) and (B), enzymes that are incorrectly annotated as non-enzymes by DomSign are excluded from the evaluation in panels (C) and (D). "Coverage" in panels (C) and (D) describes the percentage of proteins left after removal of real enzymes that were incorrectly predicted to be non-enzymes. 'Example based precision' and 'Example based recall' are used to evaluate the result as stated in Methods.

Hence, we assert that the main reason for the loss of precision and recall in DomSign was that it is too strict to differentiate enzyme candidates from protein pools. Therefore, more enzymes are mistakenly categorized into the non-enzyme group by DomSign, leading to the loss of coverage. Even though this problem causes a decrease in the "example-based precision" defined here, it does not cause errors such as predicting the wrong EC number or mistakenly identifying a non-enzyme as an enzyme. Considering that the EnzML model is difficult to implement, we posit that using DomSign would be more facile by comparison with respect to expanding the enzyme space from a large-scale dataset, as discussed in the next section.

Enzyme Prediction in UniProt-TrEMBL and KEGG

Having demonstrated the reliability of DomSign, we annotated the whole protein space to determine if we could improve the prediction coverage of enzymes with EC numbers. UniProt-TrEMBL was used in this scenario owing to its exhaustive coverage of the known protein universe. To test the precision of this enzyme prediction model, we ran the DomSign annotation against the "TrEMBL enzyme" set, which contained enzymes with single EC numbers in the TrEMBL database (6). DomSign with an 80% specificity threshold yielded a 6.6% OP ratio while assigning EC numbers to ~90% enzymes. This OP ratio, which is higher than previous validations, may be due to the greater degree of error in the TrEMBL annotation [17]. This result, combined with the performance test, demonstrated that the enzyme space expansion effort we conducted, as described below, was highly reliable.

Thus, we extended our data mining by predicting enzymes with EC numbers from all of the TrEMBL proteins. The annotation result is presented in 7. Approximately 3.9 million proteins lacking an EC number could be annotated with an EC number, and the majority of these belong to the three- or four-EC-digit group (Figure 5A). Even with a specificity threshold of 99%, the number of predicted novel enzymes was still around 3.6 million (8), further indicating the reliability of this method. By this means, we successfully raised the EC-tagged enzyme ratio from the original 12% to ~30% in TrEMBL (Figure 5A) with high precision. To further illustrate the significance of this EC resource expansion, the increased EC-tagged enzyme ratios for every genome of the bacterial taxonomy in KEGG were calculated and are presented in Figure 5B (see 9 for detailed bacterial EC number annotations in KEGG). Remarkably, on average, we raised the EC-tagged enzyme ratio of each bacterial genome from the previous 26.0% to 33.2% for 2,584 bacterial genomes in KEGG, implying that the DomSign enzyme prediction method can provide deeper insight into the metabolism of many sequenced but insufficiently characterized organisms. Taken together, DomSign enzyme predictions in TrEMBL and KEGG increased the number of EC-labeled enzymes with precision and confirmed the existence of hypothetical gaps between the real enzyme space and the functional annotation.

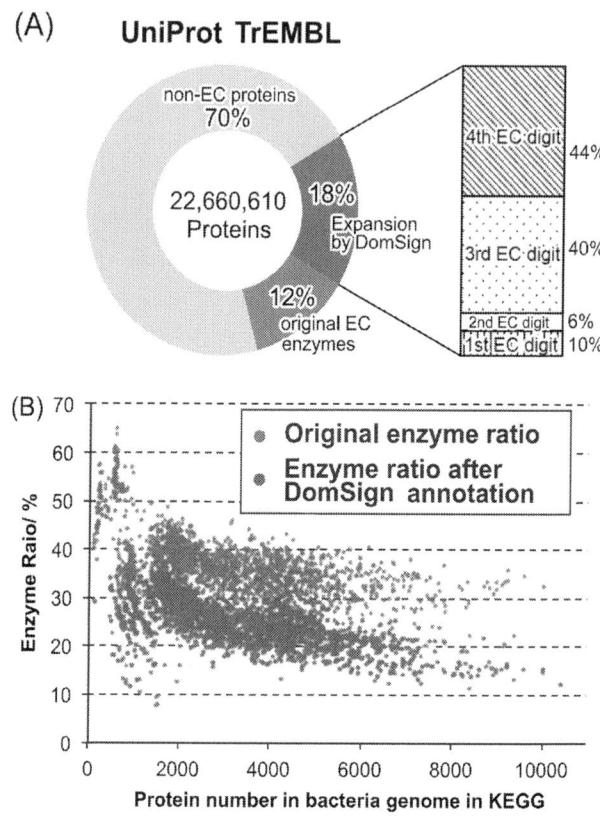

Figure 5: Expansion of enzyme space in UniProt-TrEMBL and KEGG by DomSign (specificity threshold=80%). (A) Expansion of enzyme space in UniProt-TrEMBL. The circles illustrate the distribution of three kinds of proteins in the TrEMBL database. Blue: enzymes already tagged with EC numbers in TrEMBL; red: novel enzymes exclusively predicted by DomSign; light orange: other proteins without EC numbers. The column on the right represents the ratio of EC hierarchy levels among predicted novel enzymes by DomSign. Straight line: predicted enzymes annotated as EC=x.-.-.-; blank: annotated as EC=x.x.-.-; dot: annotated as EC=x.x.x.-; slash: annotated as EC=x.x.x.x. (B) Expansion of enzyme space in KEGG. Each blue dot represents the original enzyme ratio for one particular bacteria genome in KEGG. Each red dot represents the total enzyme ratio for one particular bacteria genome after DomSign annotation. In total, 2,584 bacterial genomes were tested.

Enzyme Predictions in Metagenomic Samples

Although millions of proteins have been discovered by the biological community, our knowledge of the protein world is still far from complete, and new metagenomic data provide us with new resources to explore [13]. Thus, we chose the HMP dataset as a test set to expand the enzyme space for proteins identified in metagenomic datasets using DomSign. Additionally, a combinational annotation pipeline in HMP using BLAST, TIGRFAM and Pfam-A [24] would be expected to be a good benchmark against which to compare DomSign in the functional annotations of metagenomic sequences.

As with TrEMBL, we first applied DomSign enzyme prediction to the "HMP enzyme" set to assess DomSign's ability to predict enzymes. Compared with previous tests, much higher OP ratio (9.2%) was observed for DomSign with an 80% specificity threshold (10). Despite the inability to evaluate the reliability of HMP annotations in this analysis, similar to the high error values in automatically annotated protein datasets such as TrEMBL [17], the quality of automatic HMP annotations is probably not as high as a manually curated set like Swiss-Prot. Thus, HMP annotation errors partly explain this abnormally high OP ratio, which is strongly supported by the fact that the OP ratio reached 5.4% even for DomSign with a 99% specificity threshold. These results still support the hypothesis that the reliability of the DomSign-based enzyme space expansion in HMP metagenomic datasets is acceptable.

DomSign can recover more enzymes from this metagenomic dataset (Figure 6 and 11). Approximately one million new enzymes can be annotated with EC numbers exclusively by DomSign (around 7% of proteins in HMP set) (12), and 84% of them contain at least three EC digits. DomSign and HMP also seem to be highly complementary because half of their identified enzymes do not overlap. This is probably owing to the low Pfam-A (45.7%) coverage of HMP proteins and the appearance of many novel DSs in metagenomic sequences. The complementary properties also indicate the possibility that DomSign can detect many different catalytic functions and thus may provide further insight into the metabolic capacity of the human microbiome. To test this hypothesis, we compared the unique four-digit EC numbers retrieved by both approaches. Here, the results for DomSign with a 99% specificity threshold were used to increase the reliability of EC number assignment. As an example, 81 novel EC numbers, which were exclusively detected by DomSign with a 99% specificity threshold, were discovered from the human gut microbiome (stool sample; 13), indicating one potential biologically significant discovery. These EC numbers may reflect important components that complement the known metabolism of the human microbiome.

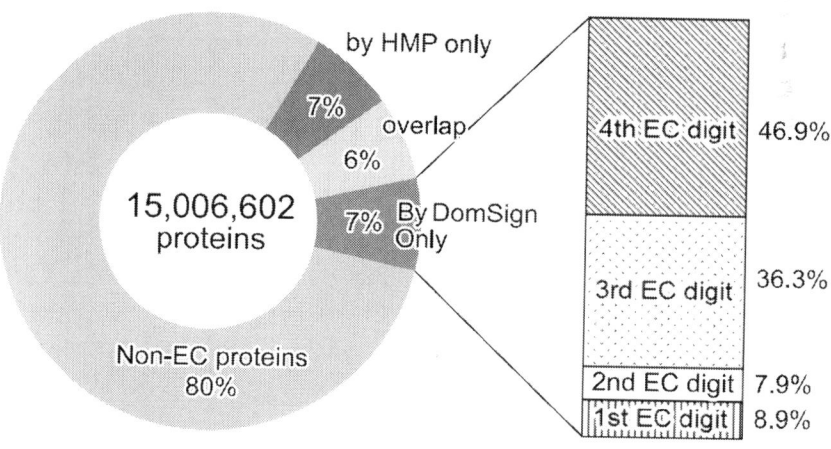

Figure 6: Expansion of enzyme space in HMP non-redundant proteins by DomSign (specificity threshold = 80%). The circles illustrate the distribution of four kinds of proteins in the HMP non-redundant dataset. Red: enzymes with EC numbers annotated exclusively by HMP; blue: novel enzymes exclusively predicted by DomSign; green: enzymes identified by both HMP and DomSign; purple: all remaining proteins. The column on the right represents the ratio of EC hierarchy levels for predicted novel enzymes by DomSign, similar to the description in Figure 5.

DISCUSSION

Limitations of DomSign

In this preliminary trial, our method performed well under diverse conditions, including having only distantly related sequences in the reference database ("sprot enzyme identity ≤ 30 %") and a query set without bias towards rich enzyme families ("SVMHL unbiased dataset"), indicating its potential to predict enzyme EC numbers in large-scale datasets. However, the precision and recall of this method are still not perfect.

First, even DomSign with a 99% specificity threshold results in a 3.6% OP ratio in the "identity $\leq 30\%$" 1,000-fold cross-validation. This is mainly because the domain architecture is unable to fully encode enzymatic activity, especially substrate specificity [38], [39]. Substrate specificity determination is complex [40], especially for some superfamilies with diverse catalytic functions [41], and thus much effort has been devoted to this task using pioneering methods such as determining key functional residues in enzymes [42], key-residue 3D templates [43] and sub strate de novo docking [44]. Future work will likely include the integration of these methodologies into our pipeline to more precisely predict the substrate specificity–determining fourth EC digit. With the development of DS databases, we can further increase the resolution of our method by involving more unique protein signatures, such as those from InterPro [34]. By this means, further increases in performance can be expected without changing the basic workflow of our method.

The comparison with SVMHL revealed variability in the performance of predicting EC number among different enzyme families. This corroborated a previous report that the worst result was obtained for oxidoreductase, as we observed with DomSign [30]. A possible solution is to utilize a combinational approach because different methodologies have diverse strengths for annotating specific enzyme families. SVMHL captures the sequence-function relationship of oxidoreductases quite well using triad abundance and structure [29]. Finally, as suggested by the comparison with EnzML, DomSign tends to have a high IA rate because it incorrectly predicts enzymes as non-enzymes. Considering that DomSign uses a very strict "yes or no" methodology to classify non-enzymes and enzymes at the first step in the pipeline, it could be improved by applying a probabilistic approach, such as the "specificity" we used in later iterations of DomSign for predicting EC numbers.

Perspective Expansion of Enzyme Space

To our knowledge, our present study represents the first systematic attempt to determine the extent to which the coverage of enzyme annotation by EC numbers could be improved, with acceptable precision, by methods beyond simple BLAST. By trying to close the gap between available EC-tagged enzymes in current databases and the real number of enzymes working in organisms, we showed that the quantity of EC-tagged enzymes can be significantly improved with high precision using relatively simple but reliable tools, such as DomSign, whether the sample is genomic or metagenomic. A series of assessments was performed to test the ability of DomSign to expand the enzyme space in large-scale protein datasets. This included a performance comparison with other benchmark enzyme annotation methods (Figures 3 and 4) and a prediction and result comparison using large-scale protein sets whose members had already been assigned EC numbers, such as TrEMBL (6) and HMP (10). Under all conditions, the precision rate was >90% and recall was quite remarkable. The results of the first large-scale critical assessment of protein function annotations (CAFA) were recently published [7]. One of the main conclusions of CAFA was that many advanced methods for protein function annotation are superior to the first generation of methods, such as BLAST. Most of the top-ranked methods in CAFA utilized a machine learning–based computational approach. As suggested by Furnham N et al. [10], however, first-generation annotation methods are still used in most research. For instance, in a previous version of SEED, an intensively used comparative genomics environment, homology-based functional transfer is the main method of annotation. This is also true for UniProt. In recent releases, UniProt incorporated the HAMAP system [45], and SEED complements its annotation strategy using a k-mer-based subsystem and FIGfam recognition approach [46]; still, these approaches depend on sequence similarity–based function transfers, such as functionally homologous family profiles. The situation is essentially the same for benchmark metagenomic projects such as HMP [24], [47]. With the development of metagenomics, many more sequences will be derived from environmental samples and will be novel compared with the current databases. In such cases, as shown in our work and that of many others [11], [13], similarity-based function transfer will struggle to achieve the desired performance.

As our work demonstrates, there is still need to improve the ability to predict more enzymes using in silico methods. Only 12% of the proteins in UniProt have EC numbers. In the HMP phase I 95% non-redundant set, this value is 13% (Figure 6). All of the values are far below the average 30% enzyme ratio of the nine intensively studied organisms [48]. We believe that a richer annotated sequence resource will result once this gap is closed using a hierarchical or top-down machine-learning method. This will allow researchers to not only study many important biological questions such as orphan enzyme gene identification [49] and metabolism network reconstruction [50] but also improve strategies used in biotechnology, including secondary metabolism gene cluster identification [51], artificial biosynthesis pathway design [52], novel enzyme mining [53] and metabolic engineering [54].

CONCLUSIONS

In this work, we developed a novel enzyme EC number prediction tool, DomSign, which is superior to conventional BLAST for the homolog unavailable scenario. In addition, other novel and outstanding enzyme functional annotation tools were selected as benchmarks and these were used to run comparisons against DomSign, which confirmed the superior or competitive ability in enzyme functional annotation of DomSign. The DomSign method requires only the amino acid sequences, without the need for existing annotations or structures. Based on the test results, the performance of DomSign should be improved by incorporating more exhaustive protein signatures, such as substrate specificity-determining residues, and revising the pipeline to select enzyme candidates using a probabilistic approach.

Using DomSign, we tried to address whether a large number of 'hidden enzymes' without EC number annotations exist in current protein databases, such as TrEMBL, KEGG and metagenomic sets like HMP. Our results preliminarily confirmed this hypothesis by significantly improving the ratio of EC-tagged enzymes in these databases. The illustration and annotation of these enzymes should significantly deepen our understanding of the metabolisms of diverse organisms or consortia, and also facilitate bioengineering by providing a richer enzyme resource. Furthermore, our results highlight the necessity to involve more advanced tools than BLAST in protein database annotations, thereby extracting more biological information from the available number of biological sequences.

AUTHORS' CONTRIBUTIONS

TW, KK, XX and TY conceived of the study. TW carried out all dataset processing and computation manipulation. TW and TY analyzed the results. HM and CZ participated in the processing of certain datasets or design of methodologies. TW and TY wrote the manuscript. All authors read and approved the final manuscript.

ACKNOWLEDGMENTS

We are grateful to Mr. Koichi Higashi and Dr. Masaaki Kotera from Tokyo Institute of Technology for their critical reading about the manuscript and constructive feedback. This work is supported by JSPS KAKENHI (Grant number 25710016) and CSC Postgraduate Scholarship Program (201306210186).

REFERENCES

1. Friedberg I: Automated protein function prediction–the genomic challenge. Brief Bioinform 2006, 7:225-42.

2. Pitkänen E, Rousu J, Ukkonen E: Computational methods for metabolic reconstruction. Curr Opin Biotechnol 2010, 21:70-7.

3. Roy A, Yang J, Zhang Y: COFACTOR: an accurate comparative algorithm for structure-based protein function annotation. Nucleic Acids Res 2012, 40(Web Server issue):W471-7.

4. Lee DA, Rentzsch R, Orengo C: GeMMA: functional subfamily classification within superfamilies of predicted protein structural domains. Nucleic Acids Res 2010, 38:720-37.

5. Gaudet P, Livstone MS, Lewis SE, Thomas PD: Phylogenetic-based propagation of functional annotations within the Gene Ontology consortium. Brief Bioinform 2011, 12:449-62.

6. Jensen LJ, Kuhn M, Stark M, Chaffron S, Creevey C, and Muller J, et al.: STRING 8–a global view on proteins and their functional interactions in 630 organisms. Nucleic Acids Res 2009, 37(Database issue):D412-6.

7. Radivojac P, Clark WT, Oron TR, Schnoes AM, Wittkop T, Sokolov A, et al.: A large-scale evaluation of computational protein function prediction. Nat Methods 2013, 10:221-7.

8. Yu C, Zavaljevski N, and Desai V, Reifman J: Genome-wide enzyme annotation with precision control: catalytic families (CatFam) databases. Proteins 2009, 74:449-60.

9. Altschul SF, Madden TL, Schäffer AA, Zhang J, and Zhang Z, and Miller W, et al.: Gapped BLAST and PSI-BLAST: a new generation of protein database search programs. Nucleic Acids Res 1997, 25:3389-402.

10. Furnham N, Garavelli JS, Apweiler R, and Thornton JM: Missing in action: enzyme functional annotations in biological databases. Nat Chem Biol 2009, 5:521-5.

11. Rost B: Enzyme function less conserved than anticipated. J Mol Biol 2002, 318:595-608.

12. Addou S, Rentzsch R, Lee D, Orengo CA: Domain-based and family-specific sequence identity thresholds increase the levels of reliable protein function transfer. J Mol Biol 2009, 387:416-30.

13. Hess M, Sczyrba A, Egan R, Kim T-W, Chokhawala H, Schroth G, et al.: Metagenomic discovery of biomass-degrading genes and genomes from cow rumen. Science 2011, 331:463-7.

14. Todd AE, Orengo CA, Thornton JM: Evolution of function in protein superfamilies, from a structural perspective. J Mol Biol 2001, 307:1113-43.

15. Shen H-B, Chou K-C: EzyPred: a top-down approach for predicting enzyme functional classes and subclasses. Biochem Biophys Res Commun 2007, 364:53-9.

16. Akiva E, Brown S, Almonacid DE, Barber AE, Custer AF, Hicks MA, et al.: The structure-function linkage database. Nucleic Acids Res 2014, 42(Database issue):D521-30.

17. Schnoes AM, Brown SD, Dodevski I, Babbitt PC: Annotation error in public databases: misannotation of molecular function in enzyme superfamilies. PLoS Comput Biol 2009, 5:e1000605.

18. Ashburner M, Ball CA, Blake JA, Botstein D, Butler H, and Cherry JM, et al.: Gene ontology: tool for the unification of biology. The gene ontology consortium. Nat Genet 2000, 25:25-9.

19. Forslund K, Sonnhammer ELL: Predicting protein function from domain content. Bioinformatics 2008, 24:1681-7.

20. Fang H, Gough J: DcGO: database of domain-centric ontologies on functions, phenotypes, diseases and more. Nucleic Acids Res 2013, 41(Database issue):D536-44.

21. Rentzsch R, Orengo CA: Protein function prediction–the power of multiplicity. Trends Biotechnol 2009, 27:210-9.

22. The UniProt Consortium: Activities at the Universal Protein Resource (UniProt) Nucleic Acids Res 2014, 42(Database issue):D191-8.

23. Kanehisa M: KEGG: kyoto encyclopedia of genes and genomes. Nucleic Acids Res 2000, 28:27-30.

24. The Human Microbiome Project Consortium: A framework for human microbiome research Nature 2012, 486:215-21.

25. Punta M, Coggill PC, Eberhardt RY, Mistry J, Tate J, Boursnell C, et al.: The Pfam protein families database. Nucleic Acids Res 2012, 40(Database issue):D290-301.

26. Messih MA, Chitale M, Bajic VB, Kihara D, Gao X: Protein domain recurrence and order can enhance prediction of protein functions. Bioinformatics 2012, 28:i444-50.

27. Eddy SR: Accelerated profile HMM searches. PLoS Comput Biol 2011, 7:e1002195.

28. Hill DP, Davis AP, Richardson JE, Corradi JP, Ringwald M, Eppig JT, et al.: Program description: strategies for biological annotation of mammalian systems: implementing gene ontologies in mouse genome informatics. Genomics 2001, 74:121-8.

29. Wang Y-C, Wang Y, Yang Z-X, and Deng N-Y: Support vector machine prediction of enzyme function with conjoint triad feature and hierarchical context. BMC Syst Biol 2011, 5 Suppl 1(Suppl 1):S6.

30. De Ferrari L, Aitken S, van Hemert J, Goryanin I: EnzML: multi-label prediction of enzyme classes using InterPro signatures. BMC Bioinformatics 2012, 13:61.

31. Tsoumakas G, Katakis I, Vlahavas I: Data Mining and Knowledge Discovery Handbook. 2010(Mlc).

32. Chou K-C: Some remarks on protein attribute prediction and pseudo amino acid composition. J Theor Biol 2011, 273:236-47.

33. Shen J, Zhang J, Luo X, Zhu W, Yu K, Chen K, et al.: Predicting protein-protein interactions based only on sequences information. Proc Natl Acad Sci U S A 2007, 104:4337-41.

34. Hunter S, Apweiler R, Attwood TK, Bairoch A, Bateman A, Binns D, et al.: InterPro: the integrative protein signature database. Nucleic Acids Res 2009, 37(Database issue):D211-5.

35. Tsoumakas G, Spyromitros-Xioufis E, Vilcek J, Vlahavas I: MULAN: a java library for multi-label learning. J Mach Learn Res 2011, 12:2411-4.

36. Desai DK, Nandi S, Srivastava PK, Lynn AM: ModEnzA: accurate identification of metabolic enzymes using function specific profile HMMs with optimised discrimination threshold and modified emission probabilities. Adv Bioinformatics 2011, 2011:743782.

37. Kumar N, Skolnick J: EFICAz2.5: application of a high-precision enzyme function predictor to 396 proteomes. Bioinformatics 2012, 28:2687-8.

38. Bashton M, Thornton JM: Domain-ligand mapping for enzymes. J Mol Recognit 2009, 23:194-208.

39. Brown SD, Gerlt JA, Seffernick JL, and Babbitt PC: A gold standard set of mechanistically diverse enzyme superfamilies. Genome Biol 2006, 7:R8.

40. Rodriguez GJ, Yao R, Lichtarge O, Wensel TG: Evolution-guided discovery and recoding of allosteric pathway specificity determinants in psychoactive bioamine receptors. Proc Natl Acad Sci U S A 2010, 107:7787-92.

41. Nagao C, Nagano N, Mizuguchi K: Relationships between functional subclasses and information contained in active-site and ligand-binding residues in diverse superfamilies. Proteins 2010, 78:2369-84.

42. Arakaki AK, Huang Y, Skolnick J: EFICAz2: enzyme function inference by a combined approach enhanced by machine learning. BMC Bioinformatics 2009, 10:107.

43. Amin SR, Erdin S, Ward RM, Lua RC, Lichtarge O: Prediction and experimental validation of enzyme substrate specificity in

protein structures. Proc Natl Acad Sci U S A 2013, 110:E4195-202.

44. Zhao S, Kumar R, and Sakai A, Vetting MW, Wood BM, Brown S, et al.: Discovery of new enzymes and metabolic pathways by using structure and genome context. Nature 2013, 502:698-702.

45. Pedruzzi I, Rivoire C, Auchincloss AH, Coudert E, Keller G, de Castro E, et al.: HAMAP in 2013, new developments in the protein family classification and annotation system. Nucleic Acids Res 2013, 41(Database issue):D584-9.

46. Overbeek R, Olson R, Pusch GD, Olsen GJ, Davis JJ, Disz T, et al.: The SEED and the rapid annotation of microbial genomes using subsystems technology (RAST). Nucleic Acids Res 2014, 42(Database issue):D206-14.

47. Tanenbaum DM, Goll J, Murphy S, Kumar P, Zafar N, Thiagarajan M, et al.: The JCVI standard operating procedure for annotating prokaryotic metagenomic shotgun sequencing data. Stand Genomic Sci 2010, 2:229-37.

48. Quester S, Schomburg D: EnzymeDetector: an integrated enzyme function prediction tool and database. BMC Bioinformatics 2011, 12:376.

49. Yamada T, Waller AS, Raes J, Zelezniak A, Perchat N, Perret A, et al.: Prediction and identification of sequences coding for orphan enzymes using genomic and metagenomic neighbours. Mol Syst Biol 2012, 8:581.

50. Orth JD, Conrad TM, Na J, Lerman JA, Nam H, Feist AM, et al.: A comprehensive genome-scale reconstruction of Escherichia coli metabolism–2011. Mol Syst Biol 2011, 7:535.

51. Medema MH, Blin K, Cimermancic P, De Jager V, Zakrzewski P, Fischbach MA, et al.:AntiSMASH: rapid identification, annotation and analysis of secondary metabolite biosynthesis gene clusters in bacterial and fungal genome sequences. Nucleic Acids Res 2011, 39(Web Server issue):W339-46.

52. Carbonell P, Parutto P, Herisson J, Pandit SB, Faulon J-L: XTMS: pathway design in an eXTended metabolic space. Nucleic Acids Res 2014, 42(Web Server issue):W389-94.

53. Schallmey M, Koopmeiners J, Wells E, Wardenga R, Schallmey A: Expanding the halohydrin dehalogenase enzyme family: identification of novel enzymes by database mining. Appl Environ Microbiol 2014, 80:7303-15.

54. Ro D-K, Paradise EM, Ouellet M, Fisher KJ, Newman KL, Ndungu JM, et al.: Production of the antimalarial drug precursor artemisinic acid in engineered yeast. Nature 2006, 440:940-3.

Development of an Artificial Neural Network Correlation for Prediction of Hold-up of Slurry Transport in Pipelines

S.K. Lahiri and K.C. Ghanta

Department of Chemical Engineering, NIT, Durgapur, West Bengal, India

ABSTRACT

In the literature, very few correlations have been proposed for hold-up prediction in slurry pipelines. However, these correlations fail to predict hold-up over a wide range of conditions. Based on a databank of around 220 measurements collected from the open literature, a correlation for hold-up was derived using artificial neural network (ANN) modeling. The hold-up for slurry was found to be a function of nine parameters such as solids concentration, particle dia, slurry velocity, pressure drop and solid and liquid properties. Statistical analysis showed that the proposed correlation has an average absolute relative error (AARE) of 2.5% and a standard deviation of 3.0%. A comparison with selected correlations in the literature showed that the developed ANN correlation noticeably improved prediction of hold-up over a wide range of operating conditions, physical properties and pipe diameters. This correlation also predicts properly the trend of the effect of the operating and design parameters on hold-up.

INTRODUCTION

Pipeline transport has been a progressive technology for conveying a large quantity of bulk materials. The modern way of pipelining prefers the concentrated slurries since hydraulic transport of dense hydro-mixtures can bring several advantages. Compared to a mechanical transport, the use of a pipeline ensures a dust free environment, demands substantially less space, makes possible full automation and requires a minimum of operating staff. On the other hand, it brings higher operational pressures and considerable demands for a high quality of pumping equipment and control system.

Power consumption costs constitute a substantial portion of operational costs for the overall pipeline transport. For that reason great attention is being paid for reduction of the hydraulic losses. The prediction of hold-up of slurries and the understanding of rheological behavior make it possible to optimize energy and water requirements.

Despite the large area of application, the available models describing the suspension mechanism do not completely satisfy engineering needs. The behavior of solids in liquid flowing through pipelines has been the subject of continuing investigation since the turn of the 19th century. From the literature (Kaushal and Tomita, 2002; Doron et al., 1987; Ghanta and Purohit, 1999; Gillies et al., 1991, Gillies et al., 1999 and Gillies and Shook, 2000; Govier and

Aziz, 1982), it is found that attempts to solve slurry flow problems may be divided into two main categories. In the first approach, one begins from the experimental facts and generalizes known correlations for some parameters by dimensional analysis, without providing an insight into the flow mechanism (e.g. Newitt et al., 1955; Zandi and Govatos, 1967; Wasp et al., 1970;Turian and Yuan, 1977, and many others). In the second approach, one starts from the basic equations of motion and numerically solves these for some situations with physical or mathematical assumptions for different terms.

Numbers of such phenomenological modeling have been proposed by Wilson, 1976 and Wilson, 1988,Wilson and Pugh (1988), Roco and Shook, 1984 and Roco and Shook, 1985, Doron et al. (1987) and many others.

Both of the above methods have their own limitations generating out of inherent complexity and poor understanding of two-phase flow systems. A predictive model with sound understanding of the fundamentals of particle laden turbulent flow, including all significant interactions, and the ability to integrate these quantitatively is not so successful till today as seen from various literatures.

To understand the hold-up phenomena it is very important to know the different flow regimes of slurry flow. There are four main flow regimes in a horizontal pipeline flow (Fig. 1). These are:

- flow with a stationary bed,
- flow with a moving bed and saltation (with or without suspension),
- heterogeneous mixture with all solids in suspension,
- pseudo-homogeneous or homogeneous mixtures with all solids in suspension.

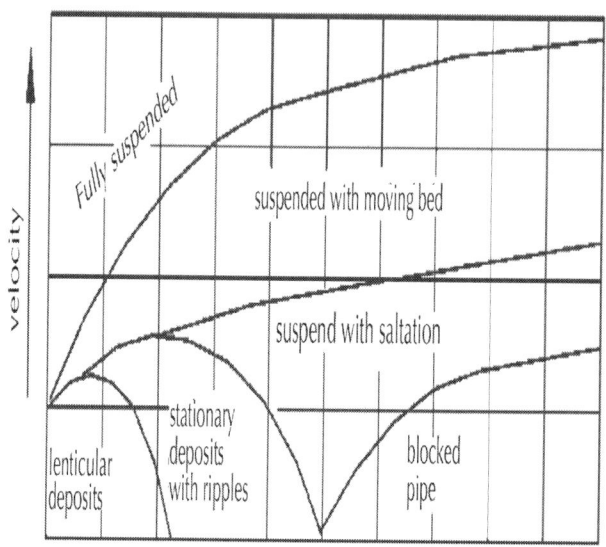

Figure 1: Flow regimes of heterogeneous flow in terms of speed versus volumetric concentration (Newitt et al., 1955).

Flow with a Stationary Bed

When the slurry flow speed is low, the bed thickens. As the fluid above the bed tries to move the solids by entrainment, they tend to roll and tumble. The particles with the lowest settling speed move as an asymmetric suspension, whereas the coarser particles build up the bed. As the speed drops even further, the pressure to maintain the flow becomes quite high and eventually the pipe blocks up. Flow with saltation and asymmetric suspension occurs above the speed of blockage. This means that the coarser particles "stand up", whereas the finer particles continue to move. Most engineering specifications require that the pipeline be designed to operate at speeds higher than those associated with saltation.

Flow with a Moving Bed

When the speed of the flow is low and there are a large number of coarse particles, the bed moves like desert sand dunes. The top particles are entrained in the moving fluid above the bed. Consequently, the upper layers of the bed move faster than the lower layers in a horizontal pipe. If the mixture were composed of a wide range of particles with different sizes and settling velocities, the bed would be composed of the particles with the highest settling speed. Particles with a moderate settling speed are maintained in an asymmetric suspension, with most particles concentrated in the lower half of the pipe, whereas the particles with the lowest settling speed move as a symmetric suspension.

Suspension Maintained by Turbulence

As the flow speed increases, turbulence is sufficient to lift more solids. All particles move in an asymmetric pattern with the coarsest at the bottom of a horizontal pipe covered with superimposed layers of medium- and fine-sized particles. Many particles may strike the bottom of the pipe and rebound. Although the flow is not symmetric, from the point of view of power consumption, this regime may be the most economical for transporting a certain mass of solids.

Symmetric Flow at High Speed

At speeds in excess of 3.3 m/s (10 ft/s), all solids may move in a symmetric pattern (but not necessarily uniformly). Sometimes, this flow is called pseudo-homogeneous because of its symmetry around the pipe axis. Power consumption is a linear relationship of the static head multiplied by the velocity, but is proportional to the cube of velocity needed to overcome friction losses. Power consumption in pseudo-homogeneous mixtures of coarse and fine particles may be excessive for long pipelines.

Hold-up

The previous section describes how different layers of solids move with different speeds, from the bottom having coarser particles, to the finer particles at the top layer in the horizontal pipe. Hold-ups are due to velocity slip of layers of particles of larger sizes, particularly in the moving bed flow pattern.

Newitt et al. (1955) conducted speed measurements of a slurry mixture in a horizontal pipe. In the case of light Plexiglas pipe, zircon or fine sand did not result in local slip; particles and water moved at the same speed. However, for coarse sand and gravel, they observed asymmetric suspension and a sliding bed. They also observed that in the upper layers of the horizontal pipe, the concentrations of larger particles were the same as for finer solids, but were marked by differences in the magnitude of the discharge rate of the lower layers.

Thus, in solid–liquid multiphase flow, the separate phases move at different average velocities and the in situ concentrations are not same as the concentrations in which the phases are introduced or removed from the system. The variation of in situ concentrations from the supply concentrations is referred to as hold-up phenomenon. The hold-up effect is measured by the hold-up ratio, given by the ratio of the average in situ concentration and mean discharge concentration:

$$\text{Hold-up ratio} = \frac{C_{\text{ins}}}{C_{\text{dis}}} = \frac{\int_A C \, dA}{\int_A (V_{m.\text{ins}}/V_{m.\text{avg}}) C \, dA}.$$

In the homogeneous suspension flow regime, it is common to assume that hold-up is negligible and therefore in situ and the input concentration are considered identical. This is not practically correct but is a suitable approximation in many cases. Hold-up has been clearly observed, especially in the flow regimes involving bed formation. For coarse particles, the difference in fluid and particle mean velocities are quite appreciable and this difference is referred to as the slip velocity. It is probable that a large contribution to the apparent slip velocity is due to fluid flowing at high velocity through the portion of low solids concentration in the pipe. Obviously, hold-up plays an important role in the failure of many empirical correlations in predicting head loss in flow regimes involving bed formation.

To facilitate the design and scale-up of pipelines and slurry pumps, there is a need for a correlation that can predict slurry hold-up over a wide range of operating conditions, physical properties and particle size distributions. Industry needs quick and easily implementable solutions. The model derived from the first principle is no doubt the best solution. But in the scenario where the basic principles of hold-up model accounting for all the interactions for slurry flow are absent, the numerical model may be promising to give some quick, easy solutions for slurry hold-up prediction.

Since the early 1980s, artificial neural networks (ANNs) have been used extensively in chemical engineering for such various applications like adaptive control, model-based control, process monitoring, fault detection, dynamic modeling and parameter estimation (Bandyopadhyay et al., 1996, Baughman and Liu, 1995, Bhatt and McAvoy, 1990, Bowen et al., 1998, Hecht-Nielsen, 1989, Schaan et al., 2000 and Venkatasubramanian and Chan, 1989). The ANN provides a non-linear mapping between input and output variables and is also useful in providing cross-correlation among these variables. The mapping is performed by the use of processing elements and connection weights. The neural network is a useful tool in rapid predictions such as steady-state or transient process flow sheet simulations, on-line process optimization and visualization and parameter estimation. Cai et al. (1994) applied Kohonen self-organizing neural networks to identify flow regimes in horizontal air–water flow. Leib et al. (1995) used a neural network model along with the mixed-cell model to predict slurry bubble column performance for the Fischer–Tropsch synthesis. Bensetiti et al. (1997), Larachi et al. (1998), Piche et al. (2001) and Iliuta et al. (2002) used an ANN to improve the prediction of various hydrodynamic parameters in packed bed and fluidized bed reactors.

Building on these studies, the focus of this work is to develop a unified correlation for overall hold-up in pipeline that can be useful for design engineers. To develop such a correlation, an approach that combines both an ANN and dimensional analysis has been used. This correlation has been derived from a broad experimental databank collected from the open literature (220 measurements covering a wide range of pipe dimensions, operating conditions and physical properties).

ANN MODELING

Neural networks are computer algorithms inspired by the way information is processed in the nervous system. An ANN is a massively parallel-distributed processor that has a natural propensity for storing experimental knowledge and making it available (Ripley, 1996 and Hornik et al., 1989). An important difference between neural networks and standard information technology (IT) solutions is their ability to learn. This learning property has yielded a new generation of algorithms. An ANN paradigm is composed of a large number of highly interconnected processing elements, analogous to neurons that are tied together with weighted connections that are analogous to synapses. Learning in biological systems involves adjustments to the synaptic connections between the neurons. This is true for ANNs as well. Learning typically occurs through training or exposure to a true set of input/output data where the training algorithm iteratively adjusts the connection weights. These connection weights represent the knowledge necessary to solve specific problems.

ANNs are being applied to an increasing number of real-world problems of considerable complexity. The advantages of an ANN-based model are:

- It can be constructed solely from the historic process input–output data (example set).
- No knowledge of the process phenomenology is necessary for the model development.
- A properly trained model possesses generalization ability due to which it can accurately predict outputs for a new input data set and even multiple input–multiple output relationships can be simultaneously approximated.

- ANN is a non-parametric estimator, making no assumptions about input distributions and uses non-linear node functions.
- ANN has natural fault tolerance due to distributed representation of information.

Network Architecture

The back propagation algorithm (BPA) assumes a feed forward neural network architecture (as shown in Fig. 2) where nodes are partitioned into layers numbered 0–L. The lowermost layer is the input layer numbered as layer 0, and the topmost layer is the output layer numbered as layer L. Back propagation addresses networks for which $L \geq 2$, containing "hidden layers" numbered 1–L-1. Hidden nodes do not directly receive inputs from nor send outputs to the external environment. Input layer nodes merely transmit input values to the hidden layer nodes and do not perform any computations. The number of input nodes equals the dimensionality of input patterns and the number of nodes in output layer is dictated by the problem under considerations. Each hidden node and output node applies the activation function to its net input. Five types of activation functions reported in literature are shown in Table 1 and Fig. 3.

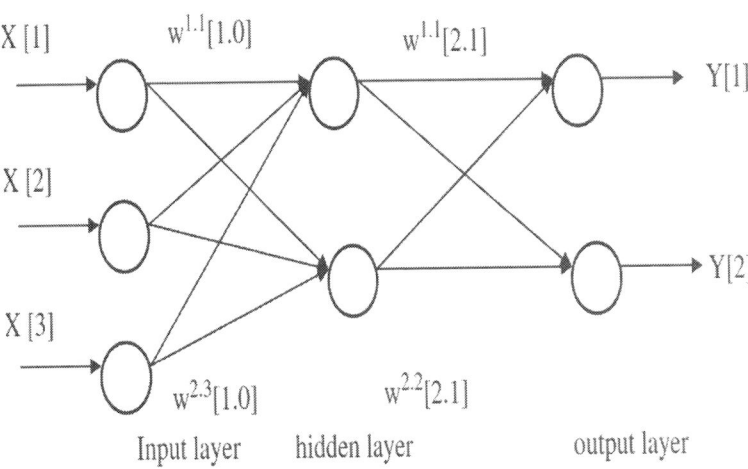

Figure 2: Architecture of feed forward network with one hidden layer.

Table 1: Different activation function

Case	Name of activation function	Equation
Case 1	logsigmoid function (logsig)	$Y_i = \dfrac{1}{(1+\exp(-net_i))}$
Case 2	tan hyperbolic function (tansig)	$Y_i = \tanh(net_i)$
Case 3	Linear function (purelin)	$Y_i = (net_i)$
Case 4	Radial basis function (radbas)	$Y_i = \exp(-neti_2)$
Case 5	Triangular basis function (tribas)	$Y_i = 1 - abs(net_i)$ if $-1 \leqslant (net_i) \leqslant 1$ $Y_i = 0$ otherwise

where $_{Yi}$ is the output from node i and n_{eti} is the input to the node $i = \sum_{wi} {}^{*}_{xi}$.

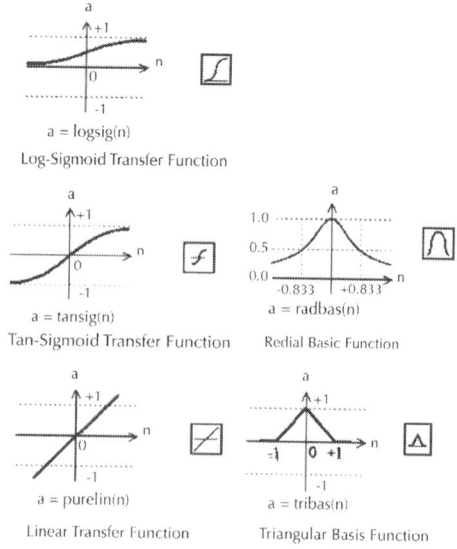

a = logsig(n)
Log-Sigmoid Transfer Function

a = tansig(n)
Tan-Sigmoid Transfer Function

a = radbas(n)
Redial Basic Function

a = purelin(n)
Linear Transfer Function

a = tribas(n)
Triangular Basis Function

Figure 3: Structure of different activation function.

Training

Training a network consists of an iterative process in which the network is given the desired inputs along with the correct outputs for those inputs. It then seeks to alter its weights to try and produce the correct output (within a reasonable error margin). If it succeeds, it has learned the training set and is ready to perform upon previously unseen data. If it fails to produce the correct output it re-reads the input and again tries to produce the correct output. The weights are slightly adjusted during each iteration through the training set (known as a training cycle) until the appropriate weights have been established. Depending upon the complexity of the task to be learned, many thousands of training cycles may be needed for the network to correctly identify the training set. Once the output is correct the weights can be used with the same network on unseen data to examine how well it performs.

Back Propagation Algorithm (BPA)

The BPA modifies network weights to minimize the mean squared error between the desired and the actual outputs of the network. Back propagation uses supervised learning in which the network is trained using data for which input as well as desired outputs are known. Once trained, the network weights are frozen and can be used to compute output values for new input samples.

The feed forward process involves presenting an input data to input layer neurons that pass the input values onto the first hidden layer. Each of the hidden layer nodes computes a weighted sum of its input and passes the sum through its activation function and presents the result to the output layer. The goal is to find a set of weights that minimize mean squared error. A typical BPA can be given as follows:

While MSE is unsatisfactory

And computational bounds are not exceeded, DO

For each input pattern $x_p, 1 \le p \ge P$,

Compute hidden node inputs

$$(\text{net}_{p,j}^{(1)}) = \sum w_{j,i}^{(1,0)} * x_{p,i};$$

Compute hidden node outputs

$$(x_{p,j}^{(1)}) = S(\sum w_{j,i}^{(1,0)} * x_{p,i});$$

Compute inputs to the output nodes

$$(\text{net}_{p,k}^{(2)}) = \sum w_{k,j}^{(2,1)} * x_{p,j}^{(1)};$$

Compute the network outputs

$$(O_{p,k}) = S(\sum w_{k,j}^{(2,1)} * x_{p,j}^{(1)});$$

Compute the error between $O_{p,k}$ and desired output $D_{p,k}$;
Modify the weights between hidden and output nodes:

$$\Delta W_{k,j}^{(2,1)} = \eta * (D_{p,k} - O_{p,k}) * S'(\text{net}_{p,k}^{(2)}) * x_{p,j}^{(1)}$$

$$(1)$$

Modify the weights between input and hidden nodes:

$$\Delta W_{j,i}^{(1,0)} = \eta * \sum \{ (D_{p,k} - O_{p,k}) * S'(\text{net}_{p,k}^{(2)}) * w_{k,j}^{(2,1)} \}$$
$$* S'(\text{net}_{p,j}^{(1)}) * x_{p,j}^{(1)}$$

$$(2)$$

End for
End while.

Different BPAs

There are several different back propagation training algorithms published in literatures. Fig. 4 shows some of those algorithms used in the present study. They have a variety of different computations and storage requirements, and no one algorithm is best suited to all locations. The basic differences between these algorithms are how they handle the weight up-gradation in (1) and (2) to reduce error and how they modify learning rate (η) to reduce convergence time.

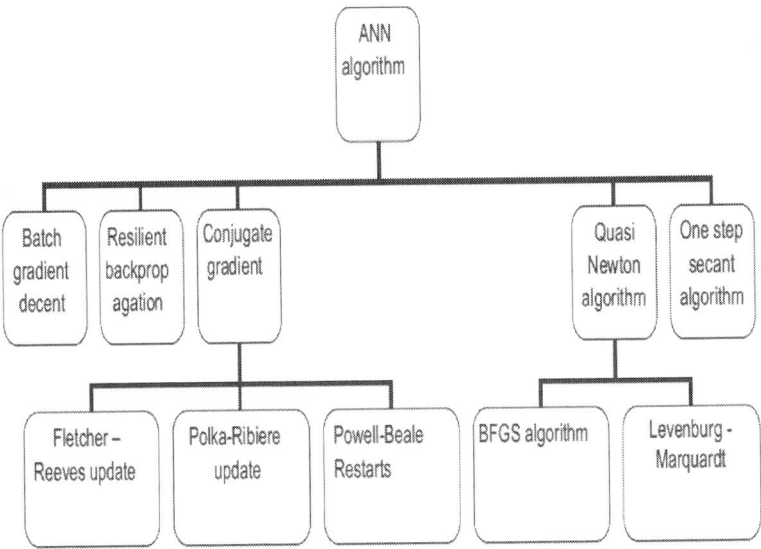

Figure 4: Different ANN algorithms published in various literatures.

The special features and comparative advantages of various algorithms as reviewed from Baughman and Liu (1995) are summarized below:

Batch Gradient Descent with Variable Learning Rate (Traingdx)

The performance of the BPA can be improved if we allow the learning rate to change during the training process. The learning rate is made responsive to the complexity of the local error surface. The learning rate is multiplied with the negative of the gradient to determine the changes to the weights and biases. The larger the learning rate, the bigger the step. If the learning rate is made too large, the algorithm becomes unstable. If the learning rate is set too small, the algorithm takes a long time to converge.

An adaptive learning rate requires some changes in the training procedure used by simple gradient descent algorithm. First, the initial network output and error are calculated. At each epoch new weights and biases are calculated using the current learning rate (typical starting value is 0.05). New outputs and errors are then calculated. If the new error exceeds the old error by more than a predefined ratio (typically 1.04), the new weights and biases are discarded. In addition, the learning rate is decreased (typically by multiplying by 0.7). Otherwise, the new weights, etc., are kept. If the new error is less than the old error, the learning rate is increased (typically by multiplying by 1.05).

This procedure increases the learning rate, but only to the extent that the network can learn without increasing large error. Thus, a near-optimal learning rate is obtained for the local terrain. When a larger learning rate could result in stable learning, the learning rate is increased. When the learning rate is too high to guarantee a decrease in error, it gets decreased until stable learning resumes.

Resilient Back propagation (Trainrp)

Multilayer networks typically use sigmoid transfer functions in the hidden layers. Sigmoid functions are characterized by the fact that their slope must approach zero as the input gets large. This causes a problem when using steepest descent to train a multilayer network with sigmoid functions, since the gradient can have a very small magnitude and, therefore, cause small changes in the weights and biases, even though the weights and biases are far from their optimal values.

The purpose of the resilient back propagation training algorithm is to eliminate these harmful effects of the magnitudes of the partial derivatives. Only the sign of the derivative is used to determine the direction of the weight update; the magnitude of the derivative has no effect on the weight update. The size of the weight change is determined by a separate update value. The update value for each weight and bias is increased by a factor whenever the derivative of the performance function with respect to that weight has the same sign for two successive iterations. The update value is decreased by another factor whenever the derivative with respect to weight changes sign from the previous iteration. If the derivative is zero, then the update value remains the same. Whenever the weights oscillate, the weight change will be reduced. If the weight continues to change in the same direction for several iterations, then the magnitude of the weight change will be increased.

Conjugate Gradient Algorithms

The basic BPA adjusts the weights in the steepest descent direction (negative of the gradient). This is the direction in which the performance function is decreasing most rapidly. It turns out that although the function decreases most rapidly along the negative of the gradient, this does not necessarily produce the fastest convergence. In the conjugate gradient algorithms a search is performed along conjugate directions, which produces generally faster convergence than steepest descent directions.

Fletcher–reeves Update (Traincgf)

All of the conjugate gradient algorithms start out by searching in the steepest descent direction (negative of the gradient) on the first iteration:

$$\mathbf{p}_0 = -\mathbf{g}_0.$$

The general procedure for determining the new search direction is to combine the new steepest descent direction with the previous search direction:

$$\mathbf{p}_k = -\mathbf{g}_k + \beta_k \mathbf{p}_{k-1}.$$

The various versions of conjugate gradient are distinguished by the manner in which the constant is computed. For the Fletcher–Reeves update the procedure is

$$\beta_k = \frac{\mathbf{g}_k^T \mathbf{g}_k}{\mathbf{g}_{k-1}^T \mathbf{g}_{k-1}}.$$

This is the ratio of the norm squared of the current gradient to the norm squared of the previous gradient.

Polak–ribiére Update (traincgp)

Polak and Ribiére proposed another version of the conjugate gradient algorithm. As with the Fletcher–Reeves algorithm, the search direction of every iteration is determined by

$$\mathbf{p}_k = -\mathbf{g}_k + \beta_k \mathbf{p}_{k-1}.$$

For the Polak–Ribiére update, the constant is computed by

$$\beta_k = \frac{\Delta \mathbf{g}_{k-1}^T \mathbf{g}_k}{\mathbf{g}_{k-1}^T \mathbf{g}_{k-1}}.$$

This is the inner product of the previous change in the gradient with the current gradient divided by the norm squared of the previous gradient.

Powell–beale Restarts (Traincgb)

For all conjugate gradient algorithms, the search direction will be periodically reset to the negative of the gradient. The standard reset point occurs when the number of iterations is equal to the number of network parameters (weights and biases), but there are other reset methods that can improve the efficiency of training. One such reset method was proposed by Powell based on an earlier version proposed by Beale. In this technique, the process will restart if there is very little orthogonality left between the current gradient and the previous gradient. This is tested with the following inequality:

$$|\mathbf{g}_{k-1}^T \mathbf{g}_k| \geqslant 0.2 \|\mathbf{g}_k\|^2.$$

If this condition is satisfied, the search direction is reset to the negative of the gradient.

Quasi-newton Algorithms

BFGS Algorithm (Trainbfg).

Newton's method is an alternative to the conjugate gradient methods for fast optimization. The basic step of Newton's method is

$$\mathbf{x}_{k+1} = \mathbf{x}_k - \mathbf{A}_k^{-1}\mathbf{g}_k,$$

where $_{Ak}$ is the Hessian matrix (second derivatives) of the performance index at the current values of the weights and biases. Newton's method often converges faster than conjugate gradient methods. Unfortunately, it is complex and expensive to compute the Hessian matrix for feed forward neural networks.

There is a class of algorithms that are based on Newton's method, which does not require calculation of second derivatives. These are called quasi-Newton (or secant) methods. They update an approximate Hessian matrix at each iteration of the algorithm. The update is computed as a function of the gradient.

Levenberg–Marquardt (Trainlm)

Like the quasi-Newton methods, the Levenberg–Marquardt algorithm was designed to approach second-order training speed without computing the Hessian matrix. When the performance function has the form of a sum of squares (as typical in training feed forward networks), then the Hessian matrix can be approximated as

$H=J^TJ$

and the gradient can be computed as

$g=J^Te,$

where **J** is the Jacobian matrix that contains first derivatives of the network errors with respect to the weights and biases and **e** is a vector of network errors. The Levenberg–Marquardt algorithm uses this approximation to the Hessian matrix in the following Newton-like update:

$$\mathbf{x}_{k+1} = \mathbf{x}_k - [\mathbf{J}^T\mathbf{J} + \mu\mathbf{I}]^{-1}\mathbf{J}^T\mathbf{e}.$$

When the scalar μ is zero, this is just Newton›s method, using the approximate Hessian matrix. When μ is large, this becomes gradient descent with a small step size. Newton's method is faster and more accurate near an error minimum. So, the aim is to shift toward Newton's method as quickly as possible. Thus, μ is decreased after each successful step (reduction in performance function) and is increased only when a tentative step would increase the performance function. In this way, the performance function will always be reduced at each iteration of the algorithm.

One Step Secant Algorithm (Trainoss)

Since the BFGS algorithm requires more storage and computation in every iteration than the conjugate gradient algorithms, there is need for a secant approximation with smaller storage and computation requirements. The one step secant (OSS) method is an attempt to bridge the gap between the conjugate gradient algorithms and the quasi-Newton (secant) algorithms. This algorithm does not store the complete Hessian matrix; it assumes that at each iteration, the previous Hessian was the identity matrix. This has the additional advantage that the new search direction can be calculated without computing a matrix inverse.

Generalizability

Neural learning is considered successful only if the system can perform well on test data on which the system has not been trained. This capability of a network is called generalizability. Given a large network, it is possible that repeated training iterations successively improve performance of the network on training data e.g. by "memorizing" training samples, but the resulting network may perform poorly on test data (unseen data). This phenomenon is called "overtraining". The proposed solution is to constantly monitor the performance of the network on the test data. Hecht-Nielsen (1989) proposes that the weight should be adjusted only on the basis of the training set, but the error should be monitored on the test set. Here, we apply the same strategy: training continues as long as the error on the test set continues to decrease and is terminated if the error on the test set increases. Training may thus be halted even if the network performance on the training set continues to improve.

DEVELOPMENT OF THE ANN-BASED CORRELATION

The development of the ANN-based correlation had been started with the collection of a large databank. The next step was to perform a neural regression, and to validate it statistically.

Collection of Data

As mentioned earlier, over the years researchers have amply quantified the hold-up of slurry flow in pipeline. In this work, about 220 experimental points have been collected from 20 sources spanning the years 1977–2000. This wide range of database includes experimental information from different physical systems to provide a unified correlation for hold-up. Table 2 indicates the wide range of the collected databank for hold-up. Table 3 shows some of these data used for neural regression.

Table 2: System and parameter studied (Doron et al., 1987, Ghanta, 1996, Ghanta and Purohit, 1999, Gillies and Shook, 2000, Kaushal and Tomita, 2002 and Roco and Shook, 1984)

Slurry system: coal water, copper ore water, sand water, gypsum water, glass water and gravel water	
Pipe diameter (m)	0.04–0.495

Particle diameter (m)*10^{-6}	38.3–13,000
Liquid density (kg/m³)	1000–1250
Solids density (kg/m³)	1370–2844
Liquid viscosity $(Pam)*10^{-3}$	0.12–4
Velocity (m/s)	1.05–4.81
Solids concentration (volume fraction)	0.0372–0.333
Max. packing concentration (volume fraction)	0.58–0.8
Pressure drop (Pa/m)	99.9–4727.7

Table 3: Some of the input and output data for neural training

	Input parameters									Output
Pipe diameter (cm)	Particle diameter (µm)	Liquid density (g/cm³)	Solids density (g/cm³)	Liquid viscosity (cp)	Max. packing concentration (volume fraction)	Velocity (m/s)	Solids concentration (volume fraction)	Pressure drop (Pa/m)		Hold-up ratio
5.26	38.30	1.00	2.33	1.00	0.69	1.11	0.1070	294.10		1.001
5.26	38.30	1.00	2.33	1.00	0.69	3.01	0.1070	1651.30		1.000
5.26	38.30	1.00	2.33	1.00	0.69	4.81	0.1070	3822.90		1.000
5.26	38.30	1.00	2.33	1.00	0.69	1.33	0.3060	542.90		1.000
5.26	38.30	1.00	2.33	1.00	0.69	3.12	0.3060	2352.60		1.000
5.26	38.30	1.00	2.33	1.00	0.69	4.70	0.3060	4727.70		1.000
20.85	190.00	1.00	1.37	1.14	0.78	2.59	0.3260	266.50		1.002
20.85	190.00	1.00	1.37	1.14	0.78	2.34	0.3270	226.30		1.003
20.85	190.00	1.00	1.37	1.14	0.78	2.01	0.3330	177.30		1.007
20.85	190.00	1.00	1.37	1.14	0.78	1.78	0.3270	147.00		1.030
20.85	190.00	1.00	1.37	1.14	0.78	1.59	0.3230	123.40		1.048
20.85	190.00	1.00	1.37	1.14	0.78	1.37	0.3270	99.90		1.064
5.15	165.00	1.00	2.65	1.00	0.58	1.66	0.0741	666.20		1.133

5.15	165.00	1.00	2.65	1.00	0.58	3.78	0.0897	2449.20	1.026
5.15	165.00	1.00	2.65	1.00	0.58	1.66	0.1694	901.30	1.104
5.15	165.00	1.00	2.65	1.00	0.58	4.17	0.1886	3428.90	1.002
5.15	165.00	1.00	2.65	1.00	0.58	1.66	0.2669	1136.40	1.049
5.15	165.00	1.00	2.65	1.00	0.58	4.33	0.2860	4408.10	1.000
26.30	165.00	1.00	2.65	1.00	0.58	2.90	0.0932	261.60	1.105
26.30	165.00	1.00	2.65	1.00	0.58	3.50	0.0921	334.10	1.086
26.30	165.00	1.00	2.65	1.00	0.58	2.90	0.1759	305.70	1.080
26.30	165.00	1.00	2.65	1.00	0.58	3.50	0.1726	382.10	1.066
26.30	165.00	1.00	2.65	1.00	0.58	2.90	0.2586	355.60	1.044
26.30	165.00	1.00	2.65	1.00	0.58	3.50	0.2597	453.60	1.032
26.30	165.00	1.00	2.65	1.00	0.58	2.90	0.3292	414.40	1.036
26.30	165.00	1.00	2.65	1.00	0.58	3.50	0.3241	526.10	1.043
49.50	165.00	1.00	2.65	1.00	0.58	3.16	0.0943	143.00	1.103
49.50	165.00	1.00	2.65	1.00	0.58	3.76	0.0923	186.10	1.083
49.50	165.00	1.00	2.65	1.00	0.58	3.07	0.1727	157.70	1.083
49.50	165.00	1.00	2.65	1.00	0.58	3.76	0.1726	210.60	1.066
49.50	165.00	1.00	2.65	1.00	0.58	3.16	0.2617	193.00	1.043
49.50	165.00	1.00	2.65	1.00	0.58	3.76	0.2602	254.70	1.034
15.85	190.00	1.00	2.65	1.30	0.58	2.50	0.1365	475.20	1.099
15.85	190.00	1.00	2.65	1.30	0.58	2.50	0.2899	630.90	1.035
15.85	190.00	1.00	2.65	0.12	0.58	3.00	0.1267	648.90	1.184
15.85	190.00	1.00	2.65	0.12	0.58	2.90	0.2775	866.70	1.081

5.07	520.00	1.00	2.65	1.00	0.70	1.90	0.0925	1175.60	1.310
5.07	520.00	1.00	2.65	1.00	0.70	2.00	0.2057	1763.40	1.202
4.00	580.00	1.25	2.27	4.00	0.65	2.88	0.1610	3926.00	1.056
4.00	580.00	1.25	2.27	4.00	0.65	2.70	0.1412	3580.00	1.133
4.00	580.00	1.25	2.27	4.00	0.65	2.01	0.1020	2217.00	1.177
4.00	580.00	1.25	2.27	4.00	0.65	1.05	0.0612	845.00	1.307
26.30	13,000.00	1.00	2.65	1.00	0.80	3.20	0.0372	842.50	2.420
26.30	13,000.00	1.00	2.65	1.00	0.80	4.00	0.0440	989.50	2.045

Identification of Input Parameters

After extensive literature survey all physical parameters that influence hold-up are put in a so-called "wish list".

Out of the number of inputs in "wish list", we used ANN regression to establish the best set of chosen inputs, which describes overall solid–liquid hold-up. The following criteria guide the choice of the set of inputs:

- The inputs should be as few as possible.
- Each input should be highly cross-correlated to the output parameter.
- These inputs should be weakly cross-correlated to each other.
- The selected input set should give the best output prediction, which is checked by using statistical analysis [e.g. average absolute relative error (AARE), standard deviation, cross-correlation coefficient].
- There should be minimum complexity in neural network architecture, i.e. a minimum number of hidden layers J.

While choosing the most expressive inputs, there is a compromise between the number of inputs and prediction.

The cross-correlation analysis which signifies the strength of the linear relation between input and output is then used to find the dependence between input and output. A number of inputs can be highly cross-correlated to output, but there should not be any strong dependency between these inputs; otherwise, it just adds to the complexity of the structure rather than contributing significantly to improve the quality of the network. One should be careful here: although the cross-correlation analysis reveals the dependence between inputs and outputs, it also hides non-monotonic relationships. This can result into losing an important input. Therefore, in this study, several sets of inputs were made and tested via rigorous trial and error on the ANN. The above-mentioned criteria were then used to identify the most pertinent set of input groups. Based on the above analysis, the input variables such as slurry velocity, pipe diameter, particle diameter, solids concentration, solid and liquid density, viscosity of flowing medium, maximum packing concentration and pressure drop have been finalized to predict hold-up in slurry pipeline. Some portion of the input and output data used for neural regression is shown in Table 2.

Neural Regression

As the magnitudes of inputs and outputs greatly differ from each other, they are normalized in 0–1 scales using following relation:

$$X_{normal} = \frac{X - X_{min}}{X_{max} - X_{min}}.$$

To avoid "overtraining" phenomena described earlier, 80% of total data set was chosen randomly for training and rest 20% was selected for validation and testing.

It has been reported that multilayer ANN models with only one hidden layer are universal approximators (Baughman and Liu, 1995). Hence, a three layer feed forward neural network (like Fig. 2) is chosen as a regression model.

As there is no previous idea about the suitability of the particular activation function, all the activation functions (listed in Table 1) are chosen in all combinations for both hidden layer and output layer. The purpose is to find out which combination gives lowest error.

The number of nodes in the hidden layer is up to the discretion of the network designer and generally depends on problem complexity. With too few nodes, the network may not be powerful enough for a given learning task. With a large number of nodes (and connections), computation is too expensive and time consuming. In the present study, the optimum number of nodes is calculated by trial-and-error method.

All the back propagation training algorithms listed in Fig. 4 were then implemented one by one to evaluate their performance with respect to validation and test error.

The statistical analysis of network prediction is based on the following performance criteria:

- The AARE should be minimum:

$$AARE = \frac{1}{N} \sum_{1}^{N} \left| \left(\frac{y_{predicted} - y_{experimental}}{y_{experimental}} \right) \right|.$$

- The standard deviation (σ) should be minimum:

$$\sigma = \sqrt{\sum_{1}^{N} \frac{1}{N-1} [|(y_{predicted_{(i)}} - y_{experimental_{(i)}})/y_{experimental_{(i)}}| - AARE]^2}.$$

- The cross-correlation coefficient (R) between input and output should be around unity:

$$R = \frac{\sum_{i=1}^{N} (y_{experimental_{(i)}} - y_{experimental_{(mean)}})(y_{predicted_{(i)}} - y_{predicted_{(mean)}})}{\sqrt{\sum_{i=1}^{N} (y_{experimental_{(i)}} - y_{experimental_{(mean)}})^2} \sqrt{\sum_{i=1}^{N} (y_{predicted_{(i)}} - y_{predicted_{(mean)}})^2}}.$$

RESULTS AND DISCUSSIONS

Optimizing Parameters in Network

After collecting the large databank, nine parameters were identified as input to ANN and the hold-up is put as target. All the ANN algorithms listed in Fig. 4 have been trained, validated and tested for best prediction using this databank.

The network parameters used for each run are summarized in Table 4.

Table 4: Detail of network

Number of input patterns used for network training	176
Number of records used for validation and testing	44

Number of input parameters	9
Number of output parameters	1
Number of hidden layers	1
Number of nodes in hidden layer	Varies between 3 and 20. Optimum number of nodes found by trial and error
Learning rate	Starting value is 0.05 and goes on changing as training proceeds
Number of neural algorithms tested	8 (Refer to Fig. 4 for names of the algorithms)
Number of activation functions tested in input layer	5
Number of activation functions tested in output layer	5

Since the prior knowledge is not there regarding the suitability of particular training algorithm, suitability of activation function at input and output layer and optimum number of nodes for solving this kind of problem, the strategy adopted here is holistic and as follows:

Step 1: Start with a training algorithm (say Levenberg–Marquardt), assume that the number of nodes in hidden layer is three, run the neural regression with all combinations of activation functions in input and output layer (total 5×5 run) and find out the AARE, standard deviation of error and cross-correlation coefficient for each run. Repeat the procedure by varying hidden layer nodes from 3 to 20 (total 18×5×5 runs for each training algorithm). Some of the results of these runs are plotted in Fig. 5, Fig. 6 and Fig. 7 for Levenberg–Marquardt algorithm.

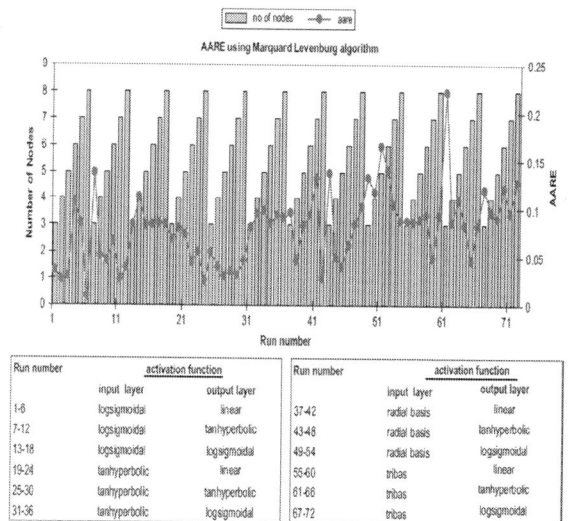

Figure 5: AARE using Marquardt–Levenberg algorithm.

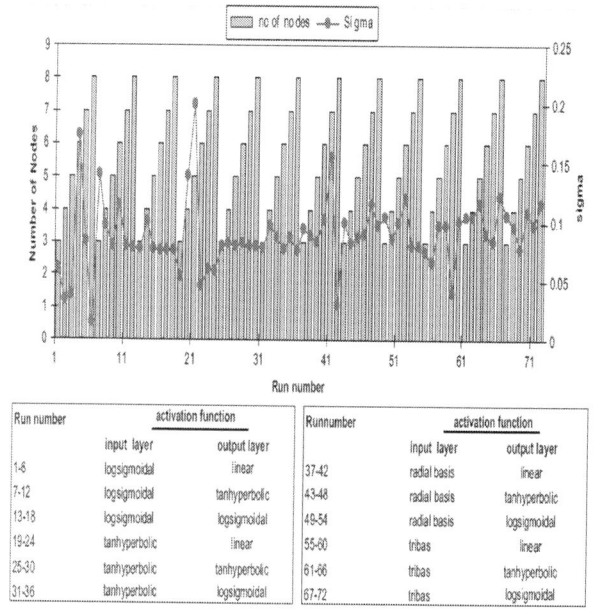

Figure 6: Standard deviation of error (σ) using Marquardt–Levenberg algorithm.

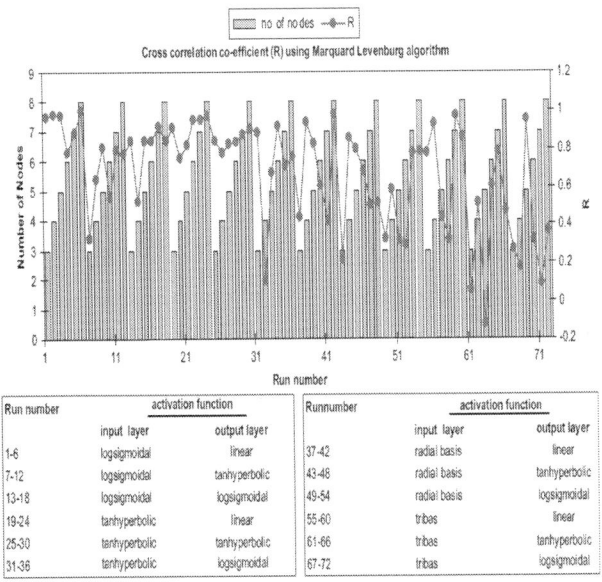

Figure 7: Cross-correlation coefficient (*R*) using Marquardt–Levenberg algorithm.

Step 2: Find out the number of nodes and input and output layer activation functions which gives lowest AARE and standard deviation of error. This may be considered as the best possible solution for that particular training algorithm and is listed in Table 5.

Table 5: Performance of different ANN algorithms

Algo-rithm	Input layer	Output layer	Number of	AARE	Standard	*R*
	transfer function	transfer function	nodes		deviation	
traincgf	logsig	purelin	6.000	0.025	0.030	0.980
trainrp	tribas	purelin	9.000	0.028	0.024	0.988
trainoss	tribas	purelin	9.000	0.033	0.048	0.970
trainlm	logsig	purelin	8.000	0.034	0.067	0.953

traincgp	logsig	purelin	10.000	0.036	0.038	0.965
trainbfg	tansig	purelin	10.000	0.041	0.048	0.965
traincgb	radbas	purelin	6.000	0.042	0.051	0.966
traingdx	tansig	purelin	8.000	0.050	0.043	0.942

Step 3: All the training algorithms are exposed with same input and output data and steps 1 and 2 are repeated for all the algorithms listed in Fig. 4 and summarized in Table 5.

For the generalization of the model, optimum iteration was found out by monitoring validation error. Training stops when validation error starts rising. Weights and learning rate at this iteration were considered for testing the suitability of the model using the test data set. A typical plot (for Fletcher–Reeves update algorithm and BFGS algorithm) in this regard has been shown in Fig. 8A and B, respectively.

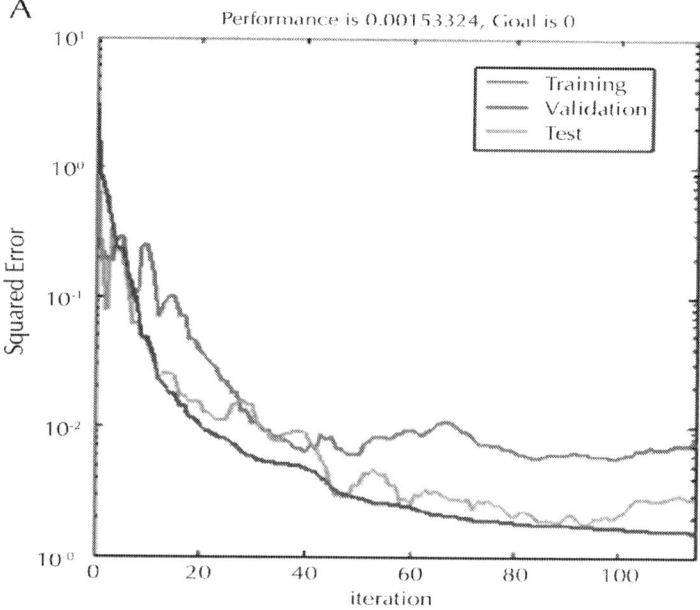

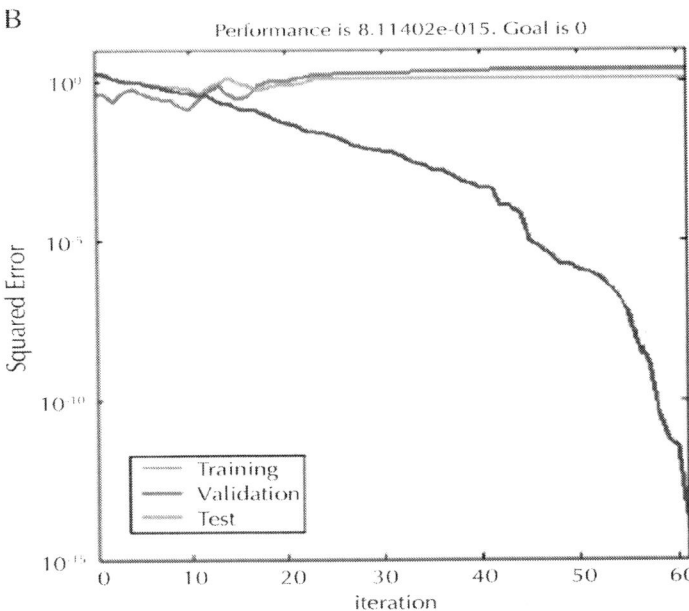

Figure 8: (A) Error reduction with iteration for Fletcher–Reeves update algorithm. (B) Error reduction with iteration for BFGS algorithm.

Effect of Number of Nodes

Effect of number of nodes on the network performance was tested for all the cited algorithms. All the combinations of input layer and output layer transfer functions were also tried along with changing of number of nodes. The optimum number of nodes for different algorithms was found out based on minimum AARE and standard deviation. A typical plot for Fletcher–Reeves update algorithm has been shown in Fig. 9.

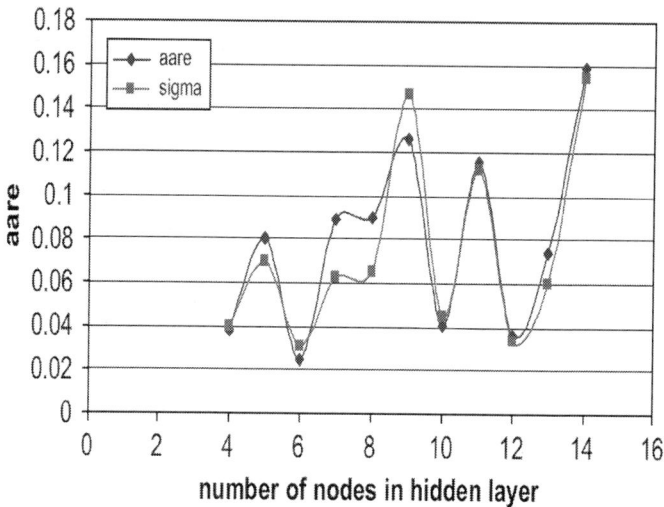

Figure 9: Effect of number of nodes in hidden layer on AARE and standard deviation of error.

The best performance of the different ANN algorithms was judged by AARE, standard deviation (σ) and cross-correlation coefficient (R). The other performance parameters, namely execution time and storage requirement, are neglected, as they are not important for this type of study. Table 5 summarizes the performance of the different algorithms. The first row of Table 5 is explained as below: When training algorithm Fletcher–Reeves update (traincgf) is applied the best result i.e., the minimum AARE (=0.025) is found with six number of nodes in hidden layer and logsigmoidal function and linear activation function at input and output layers, respectively. Same kind of explanation holds true for other rows in Table 5.

It is evident from the table that Fletcher–Reeves update algorithm with six number of nodes produces the best results among all other algorithms and predicts the hold-up with an AARE 2.5%, standard deviation 3.0% and correlation coefficient 0.98. This can be considered as a very good agreement of the model for the wide range of experimental data in comparison to the prediction by the standard available models especially for slurry flow.

A parity plot of the experimental and predicted hold-up using the ANN correlation (Fletcher–Reeves update algorithm) on the typical database has also been shown in Fig. 10. It is observed that the hold-up data are quite well fitted with the best fit curve.

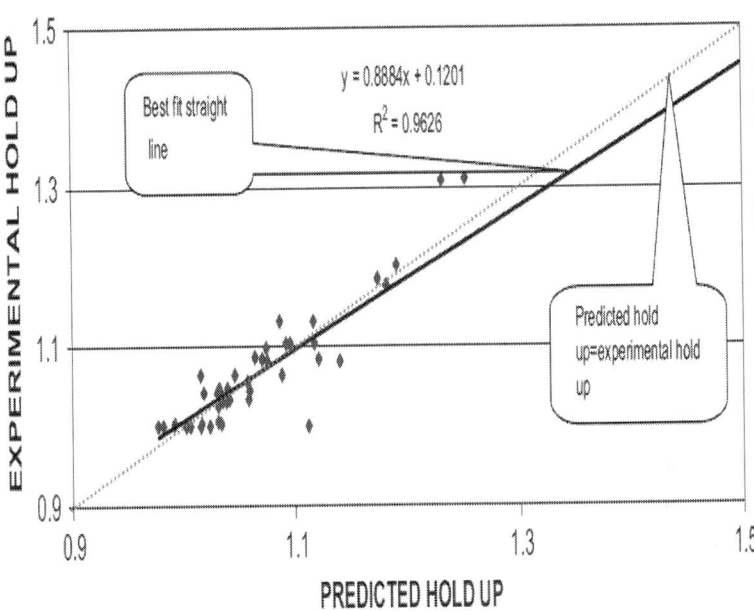

Figure 10: Parity plot for ANN correlation using the typical database.

Table 6 lists set of equations and weighting parameters for hold-up correlation.

Table 6: Set of equations and fitting parameters for neural network correlation ($i=9$, $j=6$)

$w_{i,j}$	1	2	3	4	5	6	
1	0.5471	−0.1887	0.0812	−0.1319	−0.0427	0.2064	
2	−0.3640	−0.4401	0.1656	0.9680	0.2904	0.1854	
3	0.9943	2.3325	−2.6623	−4.2112	−1.6937	0.9411	

4	−0.1299	0.5851	0.2368	−0.6478	1.0155	−0.5753	
5	−0.7460	−0.7184	−0.5126	0.3785	−0.6674	−0.7037	
6	−1.1726	1.2875	−1.7379	1.1721	−1.6191	−0.1233	
7	2.7968	−0.3366	−2.5890	2.4545	3.2347	−3.9557	
8	1.8683	−0.3992	3.2420	0.1343	2.2240	−0.5547	
9	−1.4972	−2.5752	−2.3393	−2.4976	−1.7759	0.3350	
10	−3.0206	2.4845	−2.3495	−1.7999	3.1200	4.4585	
$w_{j,k}$	1	2	3	4	5	6	7
1	−0.8403	−2.1097	2.8565	0.1720	1.0012	−0.1895	0.9710

$$H_j = \frac{1}{1 + \exp(-\sum_{i=1}^{j+1} w_{i,j} U_i)} \rightarrow S_1 = \sum_{j=1}^{j+1} w_{j,k} H_j \rightarrow \text{Hold-up ratio}$$

where $U_{10} = 1$, $H_7 = 1$ (bias)

To use the ANN correlation, these equations and parameters can be readily put in a spreadsheet file for hold-up ratio calculation for solid–liquid slurry flow.

CONCLUSIONS

The present study covers a wide range of data in slurry flow and is able to successfully predict the hold-up with reasonable accuracy (AARE 2.5%) using Fletcher–Reeves update algorithm.

Hence, ANN can be considered as an alternative solution for hold-up prediction of slurry flow where phenomenological, effective and generalized model may be difficult to develop.

REFERENCES

1.   Bandyopadhyay, J., Annamalai, S., Gauri, K.L., 1996. Application of artificial neural networks in modeling limestone-SO$_2$ reaction. A.I.Ch.E. Journal 42 (8), 2295–2302.

2.   Baughman, D.R., Liu, Y.A., 1995. Neural Networks in Bioprocessing and Chemical Engineering. Academic Press, New York.

3.   Bensetiti, Z., Larachi, F., Grandjean, B.P.A., Wild, G., 1997. Liquid saturation in cocurrent upflow fixed-bed reactors: a state-of-the-art correlation. Chemical Engineering Science 52 (21/22), 4239.

4.   Bhatt, N., McAvoy, T.J., 1990. Use of neural nets for dynamic modeling and control of chemical process systems. Computers and Chemical Engineering 14 (4), 573–582.

5.   Bowen, W.R., Jones, W.R., Yusef, H.N., 1998. Prediction of the rate of cross flow membrane ultra filtration of colloids: a neural network approach. Chemical Engineering Science 53 (22), 3793.

6.   Cai, S., Toral, H., Qui, J., Archar, J.S., 1994. Neural network based objective flow regime identification in air–water two phase flow. The Canadian Journal of Chemical Engineering 72 (3), 440.

7.   Doron, P., Granica, D., Barnea, D., 1987. Slurry flow in horizontal pipes—experimental and modeling. International Journal of Multiphase Flow 13, 535–547.

8.   Ghanta, K.C, 1996. Studies on rheological and transport characteristic of solid liquid suspension in pipeline. Ph.D. Thesis, IIT Kharagpur.

9.   Ghanta, K.C., Purohit, N.K., 1999. Pressure drop prediction in hydraulic transport of bi-dispersed particles of coal and copper ore in pipeline. The Canadian Journal of Chemical Engineering 77, 127–131.

10. Gillies, R.G., Shook, C.A., 2000. Modeling high concentration settling slurry flows. The Canadian Journal of Chemical Engineering 78, 709–716.

11. Gillies, R.G., Shook, C.A., Wilson, K.C., 1991. An improved two layer model for horizontal slurry pipeline flow. The Canadian Journal of Chemical Engineering 69, 173–178.

12. Gillies, R.G., Hill, K.B., Mckibben, M.J., Shook, C.A., 1999. Solids transport by laminar Newtonian flows. Powder Technology 104, 269–277.

13. Govier, G.W., Aziz, K., 1982. The Flow of Complex Mixtures in Pipes. Krieger, Malabar, FL.

14. Hecht-Nielsen, R., 1989. Theory of the back propagation neural network. In: Proceedings of the International Joint Conference on Neural Networks, vol. 1, pp. 593–611.

15. Hornik, K., Stinchcombe, M., White, H., 1989. Multilayer feedforward neural networks are universal approximators. Neural Networks 2, 359.

16. Iliuta, I., Grandjean, B.P.A., Larachi, F., 2002. Hydrodynamics of trickle-flow reactors: updated slip functions for the slit models. Chemical Engineering Research and Design 80 (A2), 195.

17. Kaushal, D.R., Tomita, Y., 2002. Solids concentration profiles and hold up in pipeline flow of multisized particulate slurries. International Journal of Multiphase Flow 28, 1697–1717.

18. Larachi, F., Bensetiti, Z., Grandjean, B.P.A., Wild, G., 1998. Two-phase frictional hold up in flooded-bed reactors: a state-of-the-art correlation. Chemical Engineering and Technology 21, 887.

19. Leib, T.M., Mills, P.L., Lerou, J.J., Turner, J.J., 1995. Evaluation of neural networks for simulation of three-phase bubble column reactors. Transactions of the IChemE Part A 73, 690.

20. Newitt, D.M., Richardson, J.F., Abbott, M., Turtle, R.B., 1955. Hydraulic conveying of solids in horizontal pipes. Transactions of the Institution of Chemical Engineers 33, 93–113.

21. Piche, S., Larachi, F., Grandjean, B.P.A., 2001. Improved liquid hold-up correlation for randomly packed towers. Chemical Engineering Research and Design 79 (A1), 71.

22. Ripley, B.D., 1996. Pattern Recognition and Neural Networks. Cambridge University Press, Cambridge.

23. Roco, M.C., Shook, C.A., 1984. Computational methods for coal slurry pipeline with heterogeneous size distribution. Powder Technology 39, 159–176.

24. Roco, M.C., Shook, C.A., 1985. Turbulent flow of incompressible mixtures. Journal of Fluids Engineering 107, 224–231.

25. Schaan, J., Sumner, R.J., Gillies, R.G., Shook, C.A., 2000. The effect of particle shape on pipeline friction for Newtonian slurries of fine particles. The Canadian Journal of Chemical Engineering 78, 717–725.

26. Turian, R.M., Yuan, T.F., 1977. Flow of slurries in pipelines. A.I.Ch.E. Journal 23, 232–243.

27. Venkatasubramanian, V., Chan, K., 1989. A neural network methodology for process fault diagnosis. A.I.Ch.E. Journal 35, 1993.

28. Wasp, E.J., Aude, T.C., Kenny, J.P., Seiter, R.H., Jacques, R.B., 1970. Deposition velocities, transition velocities and spatial distribution of solids in slurry pipelines. In: Proceedings of Hydro Transport, vol. 1. BHRA Fluid Engineering, Coventry, UK, pp. 53–76 (paper H4.2).

29. Wilson, K.C., 1976. A unified physically-based analysis of solid–liquid pipeline flow. In: Proceedings of the Fourth International Conference on Hydraulic Transport of Solids. BHRA Fluid Engineering, Cranfield, UK, pp. 1–16 (paper A2).

30. Wilson, K.C., 1988. Evaluation of interfacial friction for pipeline transport models. In: Proceedings of the 11th International Conference on the Hydraulic Transport of Solids in Pipes. Stratford-upon-Avon, UK, pp. 107–116 (paper B4).

31. Wilson, K.C., Pugh, F.J., 1988. Dispersive-force modeling of turbulent suspension in heterogeneous slurry flow. The Canadian Journal of Chemical Engineering 66, 721–727.

32. Zandi, I., Govatos, G., 1967. Heterogeneous flow of solids in pipelines. Proceedings of the ACSE, Journal of the Hydraulic Division 93 (HY3), 145–159.

Applicability of a Taylor–Couette Device to Characterization of Turbulent Drag Reduction in a Pipeline

Dmitry Eskin

Schlumberger DBR Technology Center, 9450-17 Avenue, Edmonton, AB, Canada T6N 1M9

ABSTRACT

A model of turbulent drag reduction in a pipe is developed. The model employs a well-known two layer representation of the boundary layer structure. An approach of Yang and Dou (2010) to model the drag reduction effect, as a phenomenon caused by a non-Newtonian rheology of a viscous sublayer flow, is employed. The modified Prandtl–Karman equation for calculation of the friction factor in a pipe flow of a dilute polymer solution is derived. This equation contains the only empirical parameter that is a function of a polymer type and concentration. The results obtained using the model developed are in a good agreement with those calculated by the Yang and Dou (2010) model, verified against experimental data. An engineering model of a turbulent dilute polymer solution flow in a Couette device is also developed. The same approach to modeling drag reduction as that in a pipe flow is applied. The model allows to compute the dimensionless torque applied to the Couette device rotor as a function of the rotation speed for a given polymer type and concentration. Thus, the empirical parameter, characterizing drag reduction by using a certain polymer additive, can be identified from laboratory Couette device experiments requiring small fluid amounts, and then applied to forecast drag reduction in industrial-scale pipeline flows.

INTRODUCTION

Drag reduction in turbulent flows, caused by polymer additives, was first reported by Toms (1948). According to the literature, addition of a small amount of high-weight polymer to fluids may reduce the pressure gradient up to 80% (Benzi et al., 2004 and L'vov et al., 2004).

In industrial applications, testing of polymer additives is associated with expensive and time-consuming experiments in flow loops for regimes, closely imitating field conditions. For example, in oil field applications drag reducers are, usually, employed for high Reynolds number flows ($Re > 10^5$). A purpose of the current work is developing a technique of characterization of drag reduction in pipelines operating under field conditions by experimental data obtained in a laboratory Couette device. A reliable model of a flow accompanied by drag reduction for both devices is needed for this purpose. This study focuses on modeling of flows in hydraulically smooth pipelines, which are most important for application of drag reducing additives. Let us first remind how a pipeline flow is modeled in engineering applications. In the absence of drag reduction agent, the velocity profile across a pipe can be expressed as (Schlichting and Gersten, 2000)

$$u^+ = y^+, \quad y^+ \leq 11.6 \tag{1}$$

$$u^+ = 2.5 \ln y^+ + 5.5, \quad y^+ > 11.6 \tag{2}$$

where $u^+ = u/u_*$, $y^+ = u_* y/v$, y is the distance from the wall, u is the streamwise velocity, $u_* = (\tau_w/\rho)^{0.5}$ is the friction velocity, v is the fluid kinematic viscosity, τ_w is the shear stress at the wall and ρ is the fluid density.

Usually, it is considered that the viscous sublayer velocity, u^+, is a linear function of y^+ only for $y^+ \leq 5$. However, the linearity is often approximately applied up to intersection of the straight line (Eq. (1)) and the logarithmic curve (Eq. (2)) at $y^+ = 11.6$. Based on the velocity distribution, described by Eqs. (1) and (2), one can derive the Prandtl–Karman equation determining the friction factor for a pipe flow as (e.g., Schlichting and Gersten, 2000)

$$\frac{1}{f^{0.5}} = 4 \log_{10}(Ref^{0.5}) - 0.4 \tag{3}$$

where $f = 2u_*^2/U^2$ is the Fanning friction factor, U is the mean flow velocity through a pipe, $Re = UD/v$ is the pipe Reynolds number and D is the pipe diameter. Note that Eq. (3) is the most accurate for high Re number flows ($Re > 10^5$).

Turbulent flows in a Couette device (see Fig. 1) have been studied in a number of papers (e.g., Lathrop et al., 1992 and Lewis and Swinney, 1999). However, there is no widely accepted engineering model of such flows. The model of Eskin (2010) is a basis for modeling a Couette device in this work. Eskin (2010) simulated a device, in which the inner cylinder rotates and the outer one is immobile. The model, based on the experimental velocity distribution in a buffer boundary layer and on the Prandtl mixing length model of turbulence in a turbulent boundary layer, establishes an analytical relation between the dimensionless torque applied to the rotor and the Reynolds number. That model and its development for a case when a drag reducing agent is present will be described in the Section 3.

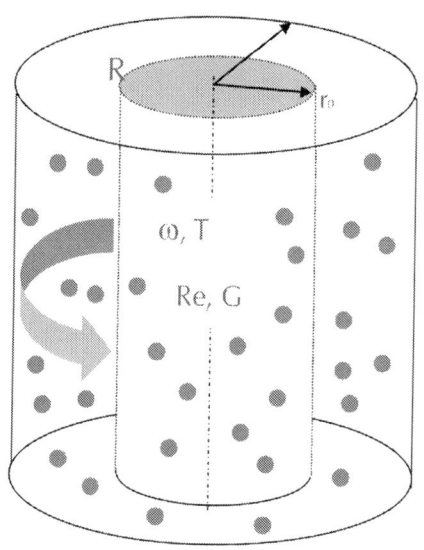

Figure 1: Couette device diagram.

In dilute polymer solution flows in pipes and channels, different investigators (e.g., Rudd, 1972, Reischman and Tiederman, 1975, Luchik and Tiederman, 1988 and Willmarth et al., 1987) studied the boundary layer structure by laser-Doppler anemometry (LDA) and particle image velocimetry (PIV). Experiments showed that adding a polymer leads to an increase in the laminar boundary layer thickness, most probably caused by a local increase in fluid viscosity, and in reducing the Reynolds shear stress in the buffer layer due to a decrease in the normal fluctuation velocity. The decrease in the Reynolds shear stress leads to the drag reduction.

According to the review of White and Mungal (2008), fundamentals of drag reduction are, usually, explained by two different theories. The first one focuses on viscous effects, whereas the second on elastic effects. The theory using the viscous explanation assumes that polymers stretching in a turbulent flow cause an increase in the effective viscosity. It is believed that in the near-wall flow polymers are stretched in the so-called buffer layer, located just above the viscous sublayer. This theory assumes that flow structure in the buffer layer causes full extension of polymers that leads to a significant increase in the elongational viscosity. It is argued that this viscosity increase suppresses turbulent fluctuations as well as causes an increase in the buffer layer thickness and a reduction in the wall friction, respectively. According to the elastic theory (Tabor and de Gennes, 1986), the elastic energy stored by partially stretched polymers determines the drag reduction effect, whereas the increase in the effective viscosity is small. The onset of drag reduction occurs when the elastic energy stored by stretched polymers reaches a level of the kinetic energy of turbulence in the buffer layer. It is believed that the smallest turbulence scale, in this case, is larger than then Kolmogorov scale and limits the turbulence spectrum from the bottom whereas the scales, which are smaller than this limiting scale, behave elastically. The elastic theory assumes that these effects also lead to thickening the buffer layer and commensurate drag reduction. Thus, according to this theory, the drag reducing properties of polymers are related to modification of coherent turbulent structures. White and Mungal (2008) also mentioned that both the viscous and the elastic theories of drag reduction onset look fundamentally different but both seem to have merit because they allow describing experimental data. Thus, in spite of a significant number of experimental data and theoretical investigations on drag reduction by polymers, the physics of this phenomenon required to be understood for its accurate simulation remains not entirely clear.

Let us mention a relatively old but important contribution into drag reduction modeling made by Virk (1971b), who suggested a semi-empirical model based on the well-known three-layer representation of the boundary layer structure (e.g., Schlichting and Gersten, 2000). According to Virk (1971b), the drag reduction caused by a polymer additive occurs due to a change in the velocity distribution across the buffer layer, while the velocity distributions across the viscous sublayer and the turbulent boundary layer are not affected by the additive. Virk (1971b) assumed that the dimensionless velocity distribution across the buffer layer ("elastic sublayer" according to the Virk terminology) is

$$u^+ = 11.7 \ln y^+ - 17 \tag{4}$$

This distribution was derived based on the empirical equation for the maximum drag reduction asymptote; i.e., the minimum friction factor achievable (Virk, 1971a). The maximum drag reduction occurs when the elastic sublayer is extended from the viscous sublayer boundary to the pipe center. Thus, according to Virk (1971b), the velocity distribution in the viscous sublayer is described by Eq. (1), in the buffer layer by Eq.(4), and in the turbulent boundary layer by an equation similar to Eq. (2), but in which the second parameter is different from 5.5 and determined as a function of a position of the of the boundary between the buffer and the turbulent boundary layers. However, this position depends on a polymer type and concentration and has to be determined experimentally. Thus, the Virk's model contains a significant empirical component. Moreover, Reischman and Tiederman (1975) measured the velocity distribution in the elastic sublayer and found it to be different from the Virk's distribution (Eq. (4)). We would like to also mention the papers of L'vov et al. (2004) and Procaccia, L'vov (2008), who successfully modeled the maximum drag reduction asymptote (Eq. (4)).

Let us now briefly describe an approach of Yang and Dou, which was employed as a basis for our model of turbulent drag reduction. These authors published a number of papers, where a technique was presented allowing rather accurate modeling of drag reduction in both smooth (Yang and Dou, 2005 and Yang and Dou, 2008) and rough (Yang and Dou, 2010) pipes. This technique is based on experimental observations (e.g., Willmarth et al., 1987 and Gyr and Tsinober, 1997), which revealed that the total shear stress in a drag reduction flow is higher than the sum of the viscous shear stress and the Reynolds shear stress if Newtonian fluid rheology is assumed. Gyr and Tsinober (1997) defined this stress deficit in a pipe flow as

$$\Gamma(y) = \tau - \left(\mu \frac{\partial u}{\partial y} + (-\rho \overline{u'v'}) \right)$$

(5)

where ρ is the fluid density, $\tau = \rho u_*^2 (1 - y/R)$ is the total shear stress, R is the channel radius, $-\rho \overline{u'v'}$ is the Reynolds shear stress, and $\mu(\partial u/\partial y)$ is the viscous shear stress.

The stress deficit effect is caused by a non-Newtonian viscoelastic rheology of a polymer additive. To take the stress deficit into account Gyr and Tsinober (1997) represented the stress deficit as follows:

$$\Gamma(y) = \rho v_{eff} \frac{du}{dy}$$

(6)

where v_{eff} is the effective viscosity.

Yang and Dou (2005) excluded the buffer layer from the model (the two-layer approach) and suggested calculating the effective viscosity across the viscous sublayer by analogy with the Boussinesq's expression for the eddy viscosity (e.g., Schlichting and Gersten, 2000), i.e. as

$$v_{eff} = \alpha_* u_* R$$

(7)

where $_{\alpha*}$ is the empirical parameter that is a function of the polymer type and its concentration.

Accounting for Eqs. (6) and (7) one can write Eq. (5) in the following form (Yang and Dou, 2008):

$$u_*^2 \left(1 - \frac{y}{R} \right) = (v + v_{eff}) \frac{du}{dy} - \overline{u'v'} = v D_* \frac{du}{dy} - \overline{u'v'}$$

(8)

where $_{D*}$ is the drag reduction (DR) parameter that is calculated as

$$D_* = 1 + \alpha_* \frac{u_* R}{v} = 1 + \alpha_* R^+$$

(9)

$$R^+ = 0.5\mathrm{Re}\sqrt{\frac{f}{2}}$$

where   is the dimensionless pipe radius equal to the Reynolds number based on the friction velocity. In the dimensionless form Eq. (8) can be written as

$$\left(1 - \frac{y^+}{R^+}\right) = D_* \frac{du^+}{dy^+} - \overline{u'^+ v'^+}$$

(10)

The boundary condition for Eq. (10) is zero velocity at the wall: $u^+(0)=0$.

Using the stochastic theory of turbulence Yang and Dou (2008) modeled the Reynolds shear stress. Substituting this stress into Eq. (10) they obtained the equation for the dimensionless velocity distribution $u^+(y^+)$ as

$$u^+ = \frac{1}{\kappa}\ln\left(1 + \frac{\kappa y^+}{2D_*}\right) + \frac{1}{2}\left(\frac{\delta^+}{D_*} + \frac{1}{\kappa}\right)\left(\frac{\kappa y^+}{2D_* + \kappa y^+}\right)^2 + \frac{1}{\kappa}\frac{\kappa y^+}{2D_* + \kappa y^+}$$

(11)

where κ=0.4 is the von Karman constant, and δ^+ is the dimensionless viscous sublayer thickness.

The key element of the Yang and Dou (2008) model is the dimensionless viscous sublayer thickness that is calculated by the experimental correlation (Yang and Dou, 2005):

$$\delta^+ = 11.6\, D_*^3$$

(12)

Based on the velocity distribution, the friction factor is determined straightforwardly. The dimensionless mean flow velocity is calculated by averaging the velocity u^+ over the pipe cross-section as

$$\frac{U}{u_*} = \sqrt{\frac{2}{f}} = \frac{1}{\pi R^{+2}}\int_0^R u^+ 2\pi R^+ (R^+ - y^+)dy^+$$

(13)

After integration of Eq. (13) and further simplifications, Yang and Dou (2010) obtained the equation for the friction factor that they recommended to use for engineering calculations. The final form of this equation is (Yang and Dou, 2010)

$$\sqrt{\frac{2}{f}} = 2.5 \ln \frac{R^+}{D_*} - 6.69 \left(\frac{R^+}{D_*^{3.5}} \right)^{-0.72} + 5.8D_*^2 - 4$$

(14)

It was demonstrated (Yang and Dou, 2010) by comparing the calculation results with the experimental data of Virk (1971a) that Eq. (14) accurately predicts the friction factor if the empirical coefficient α_* defining D^* is identified correctly. For practical purposes, Eq. (14) can be applied as follows. For a given additive type and concentration a several experiments are performed with an experimental flow loop. Each experiment is carried out at a different flow rate. Then, the coefficient α_* is identified to provide the best fit of the computational results to the experimental data.

Thus, in contrast to the Virk's approach (Virk, 1971b), the described technique assumes that the drag reduction is caused by the non-Newtonian behavior of the laminar boundary layer. The thickness of this layer rapidly increases with the polymer additive concentration that leads to a reduction in the Reynolds shear stress in the turbulence boundary layer (Yang and Dou, 2010). This modeling approach can be classified as a modification of the viscous theory of drag reduction (e.g., White and Mungal, 2008) briefly described above. However, in contrast to the standard viscous theory, the viscosity increase caused by polymer stretching occurs in the viscous sublayer but not in the buffer layer that is excluded from the model.

We would like to emphasize that Taylor–Couette apparatuses have been used for studies of turbulent drag reduction in the past (e.g., Kalashnikov, 1998 and Koeltzsch et al., 2003). However, no engineering model of drag reduction phenomenon in a Taylor–Couette flow like that in a pipe flow (e.g., Yang and Dou, 2010) has ever been presented in open literature. Investigators either limited their modeling efforts to engineering evaluations as Kalashnikov (1998) and Koeltzsch et al. (2003) did, or modeled drag reduction using sophisticated numerical methods, which can hardly be used for engineering calculations (e.g., Liu and Khomami, 2013). Kalashnikov (1998) studied drag reduction in a Taylor–Couette device that included a rotating outer cylinder and an immobile inner cylinder. Although, this author suggested dimensionless criteria for drag reduction characterization, they have a limited applicability. The results of Kalashnikov (1998) cannot be used in practical cases because owing to a small height of the experimental device, the friction losses, caused by viscous friction in clearances between side surfaces of the rotor and the stator, were comparable to the friction losses in a gap between the cylindrical surfaces. Therefore, an accurate interpretation of the experimental results was impossible. Koeltzsch et al. (2003) studied turbulent drag reduction in a Taylor–Couette device of a similar design and performed simple engineering estimations to analyze experimental data obtained. Recently, Liu and Khomami (2013) modeled a polymer solution flow in a Taylor–Couette device using direct numerical simulations (DNS). They discovered the polymer-induced breakup of large Taylor vortexes into small ones, which led to a significant increase of drag forces at the walls. However, this effect was detected at a relatively low Reynolds number (Rec=5000) and cannot occur in a flow regime, which need to be provided in a Taylor–Couette device to characterize drag reduction in a pipeline flow under field conditions (see the explanations in the Section 3).

MODELING DRAG REDUCTION IN A PIPE FLOW

Initially, our idea was to directly use an approach of Yang and Dou for modeling a Couette flow. However, their turbulent viscosity model (Yang and Dou, 2008) cannot be applied to a Couette flow straightforwardly. That model gives zero turbulent viscosity in the pipe center that contradicts to experimental observations (e.g., Schlichting and Gersten, 2000). Thus, although the Yang and Dou model provides rather accurate velocity profiles in a pipe flow, it cannot be applied to a Couette flow that is characterized by high turbulence intensity at the gap centerline. Because Virk (1971a) found out experimentally that in the turbulent core of a polymer solution pipe flow the velocity distribution can be described by the same equation as that used for a Newtonian flow (see Eqs.(1)and (2)) but with some velocity increment, ΔB, we can use this fact in our model. We will look for the dimensionless velocity profile in a pipe in the following form (Virk 1971a):

$$u^+ = 2.5 \ln y^+ + 5.5 + \Delta B \qquad (15)$$

Let us employ the known modeling approach (Eqs. (1) and (2)) assuming the two-layer structure of the boundary layer. According to it, the viscous sublayer, characterized by a linear velocity distribution, is extended (up to $\delta^+=11.6$ for a Newtonian flow) whereas the buffer layer is ignored. For description of the viscous sublayer we will use the approach of Yang and Dou, described above. The momentum equation for the viscous sublayer follows from Eq. (8) and takes the following form:

$$D_* \frac{du^+}{dy^+} = 1 \qquad (16)$$

Then, the velocity distribution across this sublayer is

$$u^+ = \frac{y^+}{D_*}, \; y^+ \leq \delta^+ \qquad (17)$$

The thickness of the viscous sublayer for a polymer solution flow is calculated by Eq. (12). Then, the velocity at the boundary separating the viscous sublayer and the turbulent boundary layer is

$$u^+(\delta^+) = \frac{\delta^+}{D_*} = \frac{11.6 D_*^3}{D_*} = 11.6 D_*^2 \qquad (18)$$

Equating the velocity expressed by Eq. (18) to that determined by Eq. (15) we obtain the equation for the velocity increment B as

$$\Delta B = 11.6(D_*^2 - 1) - 7.5 \ln D_* \qquad (19)$$

Then, the velocity distribution across the turbulent boundary layer is:

$$u^+ = 2.5 \ln y^+ + 11.6D_*^2 - 7.5 \ln D_* - 6.1$$

$$(20)$$

The equation for the Fanning friction factor is derived by averaging this distribution over the pipe cross-section (see Eq. (13)). As a result, we obtain the Prandtl–Karman equation modified for polymer solutions as follows:

$$\frac{1}{f^{0.5}} = 4 \log_{10}(Ref^{0.5}) + 8.2D_*^2 - 8.6 - 12.2 \log_{10} D_*$$

$$(21)$$

As one can see, this equation is reduced to the regular Prandtl–Karman equation (Eq. (3)) if the DR parameter $_D$ is 1. Since Eq. (21) is based on the same turbulence model as Eq. (3), we expect that Eq.(21) provides the highest accuracy at high Reynolds numbers ($Re > 10^5$).

To validate the model developed we compared it with the model of Yang and Dou (Eq. (14)). The authors (Yang and Dou, 2010) demonstrated a very good agreement of their calculation results with the experimental data of Virk (1971a). In Fig. 2 one can see the curves $1/f^{0.5}$ vs. $Ref^{0.5} = 2\sqrt{2R^+}$, calculated by both the models for the different water-based polymer solution flows. The polymer type and its corresponding concentration for each curve are also shown in Fig. 2. The solid lines correspond to Eq. (21), the dash-point lines to Eq. (14). Yang and Dou (2010) identified the coefficients α_* from the experimental data of Virk (1971a). We identified the coefficients for our model (α_{*_o}) to provide the best fit of our results to the curves calculated by Eq. (14). Both the coefficients α_* and α_{*_o} are also given in Fig. 2. One can see that the agreement between the results obtained by the different models is reasonably good. The differences between the curves practically disappear at relatively high Reynolds numbers. This observation is in line with our expectations of the highest accuracy of the modified Prandtl–Karman equation (Eq. (21)) for high Re numbers. Also note that the fitting coefficients α_* and α_{*_o} employed by the compared models are noticeably different.

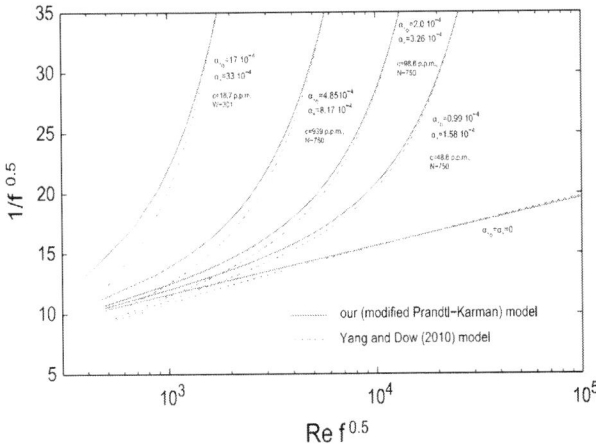

Figure 2: Comparison of the drag reduction effect in a pipe flow, estimated by the model of Yang and Dou (2010), with the effect obtained by the modified Prandtl–Karman model.

The comparison of our model with the model of Yang and Dou (2010) showed that the model developed can be employed for modeling turbulent drag reduction in pipelines of industrial scales, which are usually characterized by high Reynolds numbers. The successful modeling of turbulent drag reduction in a pipe flow by the modified Prandtl–Karman model shows the suitability of the Yang and Dou approach to model a Taylor–Couette flow. This is because the Prandtl mixing length turbulence model that is the basis of the Prandtl–Karman model is suitable for description of a Taylor–Couette flow (Schlichting and Gersten, 2000 and Eskin, 2010).

TAYLOR–COUETTE DEVICE

We will consider a Couette device where the inner cylinder rotates whereas the outer one is immobile (seeFig. 1). It is convenient to study a flow in such a device in terms of the dimensionless torque and the Reynolds number. The dimensionless torque is defined as follows:

$$G = \frac{T}{\rho v^2 L}$$

(22)

where L is the Couette device height.

The Reynolds number for a Couette flow is calculated as

$$Re_c = \frac{\omega r_0 (R - r_0)}{v}$$

(23)

where R is the outer cylinder radius, r_0 is the inner cylinder radius, $T = \tau_w 2\pi R^2 L$, and ω is the rotor angular velocity.

As it was already mentioned in the Section 1, Eskin (2010) developed a model of a flow in a Couette device. He employed an empirical velocity distribution across both the viscous and the buffer boundary layers (Schlichting and Gersten, 2000) and the Prandtl mixing length turbulence model for a turbulent core flow. Because the Couette flow model needed to be applicable for a wide gap Couette device, the streamline curvature was taken into account within the turbulence model. The author (Eskin, 2010) derived an analytical equation linking the dimensionless torque and the Couette device Reynolds number as

$$\varsigma(\eta) \frac{Re_c}{\sqrt{G}} = \ln \sqrt{G} + \phi(\eta) + \xi$$

(24)

where $\eta = r_0/R$, $\varsigma(\eta) = k / \left((1/\eta) + \eta \right) \left(\sqrt{2\pi/1 - \eta} \right)$, $\phi = \left(2/(1/\eta) + \eta \right) - \ln\left(1 + \eta/1 - \eta \right)$, $\kappa = 0.44$ is the von-Karman constant for a Couette flow (Lewis and Swinney, 1999) and ξ is the parameter.

The parameter is a function of the boundary (initial in this case) condition for a flow domain, for which the Prandtl mixing length model is applied. This parameter depends on the total viscous and buffer boundary layer thickness and the velocity distribution across this thickness, and calculated as (Eskin, 2010):

$$\xi = \kappa\lambda - (1 + \ln(b)) - \ln\sqrt{2\pi}$$

(25)

where $b = \delta_{tot}^+$ is the total dimensionless thickness of the buffer and the viscous boundary layers, and $\lambda = u^+(b)$ is the dimensionless velocity at the top of the layer of the thickness b.

Based on the experimental dimensionless velocity distribution across the buffer layer of a Couette flow, Eskin (2010) using the data of Schlichting and Gersten (2000) assumed b=70 that provides =14.94, and then Eq. (23) gives =0.406.

Note that Eq. (24) was derived assuming that the total thickness of the viscous and the buffer boundary layers, b, is negligibly small, compared to the wall radius. Eq. (24) demonstrated a reasonably good accuracy for relatively high Reynolds numbers (R_{ec}>13,000), at which an effect of Taylor vortices on a turbulent flow structure is negligibly small (Lewis and Swinney, 1999).

Flow Model

To model the turbulent drag reduction effect in a Taylor–Couette device by the same approach as that employed for modeling a pipe flow, we need to modify the Couette flow model (Eq. (24)) applying the two sublayers approach to flow field description. Also, because our estimations made for a pipe flow show that for a polymer solution case the viscous sublayer thickness may reach relatively high values, we discarded the assumption about smallness of the viscous sublayer thickness in comparison to the wall radius.

The velocity distribution across the viscous sublayer in a Couette flow is also linear:

$$u^+ = y^+, \quad y^+ \leq \delta^+$$

(26)

By analogy with the pipe flow, the thickness of the viscous sublayer can be determined at the intersection of the linear velocity profile within the viscous sublayer and the velocity profile in the turbulent core. The velocity distribution across the turbulent region, confined by the outer boundary of the viscous sublayer at the outer wall and the gap centerline, is described by the ordinary differential equation as (Eskin, 2010)

$$\frac{d(u/r)}{dr} = -\frac{u_{0*}R}{\kappa(R-r)r^2}$$

(27)

The boundary condition for this equation is the velocity at the

viscous sublayer surface $u^+(\delta_0^+) = \lambda$. The analytical solution of Eq. (27) is

$$\frac{u(r)}{u_{0*}} = \frac{1}{\kappa}\left(1 + \frac{r}{R}\ln\left(\frac{R}{r}-1\right)\right) + \gamma\frac{r}{R},$$

(28)

where

$$\gamma = \frac{\lambda - 1/\kappa(1+(1-a)\ln(a/1-a))}{1-a}$$

(29)

and $a = \delta_0^+/R^+$, $\delta_0^+ = \delta_0/(v/u_{0*})$ is the dimensionless viscous sublayer thickness at the outer wall.

The dimensionless outer Couette device radius R^+ is expressed through the dimensionless torque G as

$$R^+ = \sqrt{\frac{G}{2\pi}}$$

(30)

The momentum conservation equation for the turbulent flow region between the centerline and the outer surface of the viscous sublayer at the inner cylinder is (Eskin, 2010)

$$\frac{d(u/r)}{dr} = -\frac{u_{0*}R}{\kappa(r-r_0)r^2}$$

(31)

The initial condition for this equation is the velocity at the gap centerline that is calculated using Eq. (28) as

$$\frac{u(R_m)}{u_{0*}} = \frac{1}{\kappa}\left(1 + \frac{1+\eta}{2}\ln\left(\frac{1-\eta}{1+\eta}\right)\right) + \gamma\frac{1+\eta}{2} \tag{32}$$

where $_{Rm} = 0.5(_{r0}+R)$.

Then, the analytical solution of Eq. (31) is (Eskin, 2010)

$$\frac{u(r)}{u_{0*}} = \frac{u(R_m)}{u_{0*}}\frac{r}{R_m} + \frac{1}{\kappa}\frac{R}{r_0}\left(-1 + \frac{r}{R_m} + \frac{r}{r_0}\ln\left(\frac{1-(r_0/R_m)}{1-(r_0/r)}\right)\right) \tag{33}$$

The circumferential velocity of the inner cylinder can be calculated as (Eskin, 2010)

$$U_i = u(r_0 + \delta_i) + \lambda u_{i*} \tag{34}$$

where u_{i*} is the friction velocity at the inner wall, and δ_i is the viscous sublayer thickness at the inner wall.

One can easily express the parameters of the boundary layer at the inner cylinder through those at the outer cylinder (see Eskin, 2010) and rewrite Eq. (34) as follows:

$$\frac{U_i}{u_{0*}} = \frac{u(r_0 + \delta_i)}{u_{0*}} + \frac{\lambda}{\eta} \tag{35}$$

Expressing the left-hand side of this equation through the dimensionless torque and the Reynolds number we obtain

$$\frac{(2\pi)^{0.5}}{1-\eta}\frac{Re_c}{G^{0.5}} = \frac{u(r_0 + \delta_i)}{u_{0*}} + \frac{\lambda}{\eta} \tag{36}$$

The velocity $u(_{r0}+_{\delta i})=u(_{r0}+_{\delta0}\eta)$ is calculated by Eq. (33), and after performing a routine math takes the form:

$$\frac{u(r_0 + \delta_i)}{u_{0*}} = \frac{u(R_m)}{u_{0*}}\frac{2\eta}{1+\eta}(1+a) + \frac{1}{\kappa}\left(-\frac{1}{\eta} + \frac{2(1+a)}{1+\eta} + \frac{1+a}{\eta}\ln\frac{(1-\eta)(1+a)}{(1+\eta)\,a}\right) \tag{37}$$

The set of Eqs. (32), (36) and (37) allows calculating the dimensionless torque applied to the Couette device rotor. Note that we employed the von-Karman constant $=0.44$, identified by Lewis and Swinney (1999) from the Couette flow experimental data, and considering similarity of the near-wall boundary layer in a Couette flow to that in a pipe flow, assumed $\delta_0^+ = 11.6$ Following the two-layer approach to modeling the boundary layer, employed for a pipe flow, we consider $\delta_0^+ = 11.6$ and, therefore, the velocity at the viscous sublayer boundary is determined by Eq. (26) as $u^+(11.6) = = 11.6$. We also checked how the model developed performs against the simplified model (Eq. (24)) that neglects the boundary layer thickness in comparison with the Couette device radius. It turned out that the solutions obtained by the developed and the simplified models are practically identical for $R_{ec} > 13,000$. This result was expected, because in the case of a pure fluid (free of a drag reducer) the viscous sublayer thickness is negligibly small in comparison with the Couette device radii.

In Fig. 3 we showed the relative deviations of the dimensionless torque calculated by Eq. (36) from the data of Lewis and Swinney (1999), who accurately approximated their experimental results by the following correlation:

$$\log_{10} G = -0.00636(\log_{10} Re_c)^3 + 0.1349(\log_{10} Re_c)^2 + 0.885(\log_{10} Re_c) + 1.61 \qquad (38)$$

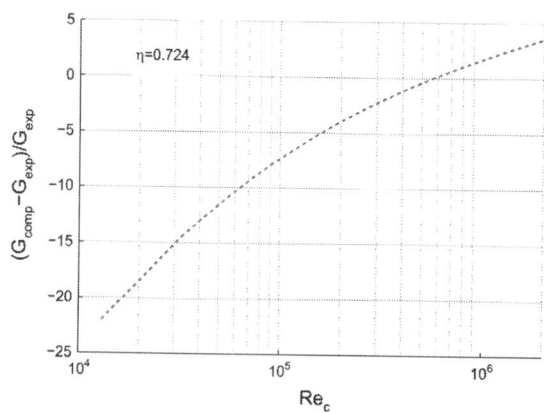

Figure 3: Relative deviation of the dimensionless torque computed by the model developed from the experimental data vs. Reynolds number.

One can see that the agreement is very good in the wide range of Rec numbers, except the relatively low Rec values. The accuracy is excellent for $Rec > 10^5$. For such high Rec numbers the deviations of the computed results from the measured data do not exceed 7%.

Drag Reduction Model

Application of the approach proposed for modeling turbulent drag reduction in a pipe to a Couette flow is straightforward. The drag reduction parameters are different for the outer and the inner cylinders and calculated as

$$D_{0*} = 1 + \alpha_* \frac{u_{0*}H}{2\nu} \tag{39}$$

$$D_{i*} = 1 + \alpha_* \frac{u_{i*}H}{2\nu} \tag{40}$$

where H is the Couette device gap width.

By taking into account Eq. (36), one can rewrite Eqs. (39) and (40) as follows:

$$D_{0*} = 1 + \alpha_* \frac{Re_c}{2} \frac{u_{0*}}{U_i} = 1 + \frac{\alpha_*}{2}(1-\eta)\sqrt{\frac{G}{2\pi}} \tag{41}$$

$$D_{i*} = 1 + \alpha_* \frac{Re_c}{2} \frac{u_{i*}}{U_i} = 1 + \frac{\alpha_*(1-\eta)}{2} \frac{1}{\eta}\sqrt{\frac{G}{2\pi}} \tag{42}$$

By analogy with a pipe flow, the viscous sublayer thicknesses at the outer and the inner surfaces of a Couette device are

$$\delta_0^+ = 11.6\, D_{0*}^3 \tag{43}$$

$$\delta_i^+ = 11.6\, D_{i*}^3 \tag{44}$$

Then, the corresponding flow velocities at the top of the viscous sublayers, needed for application of the Couette flow model, are

$$\lambda_0 = \frac{\delta_0^+}{D_{0*}} = 11.6 D_{0*}^2$$

(45)

$$\lambda_i = \frac{\delta_i^+}{D_{i*}} = 11.6 D_{i*}^2$$

(46)

Let us briefly formulate a computational algorithm of a polymer solution flow in a Couette device. Eq. (36) is the major model equation that has to be solved by iterations over the Re_c number. The dimensionless velocity $\lambda = \lambda_i$ is determined by Eq. (46). The flow velocity at the inner viscous sublayer surface is calculated by the equation derived on the basis of Eq. (33) and has the following form:

$$\frac{u(r_0 + \delta_i)}{u_{0*}} = \frac{u(R_m)}{u_{0*}} \frac{2\eta}{1+\eta}(1+m) + \frac{1}{\kappa}\left(-\frac{1}{\eta} + \frac{2(1+m)}{1+\eta} + \frac{1+m}{\eta}\ln\frac{(1-\eta)(1+m)}{(1+\eta)}\frac{m}{m}\right)$$

(47)

where $m = \delta_i^+ / r_0^+$, $r_0^+ = R^+$ (R^+ is calculated by Eq. (30)).

The dimensionless velocity at the gap centerline $u(_{Rm})/u_{0*}$ is computed by Eq. (32). The parameter γ is calculated by Eq. (29). The parameters a and $\lambda = \lambda_0$, needed for using this equation, are determined for the outer cylinder by Eqs. (41), (43) and (45).

Computational Examples and Discussion of the Drag Reduction Model in a Taylor–Couette Device

The best way to illustrate the model performance is to present the calculation results in the same coordinates, which were used for a pipe flow (see Fig. 2). Because the Couette device Reynolds number is based on the rotor circumferential velocity, we will employ flow parameters at the inner cylinder surface for further analysis. The shear stress at the inner cylinder wall can be calculated as:

$$\tau_{wi} = \frac{\rho f U_i^2}{2}$$

(48)

where f is the Fanning friction factor for a rotating wall.

From Eqs. (23), (24), (25), (26), (27), (28), (29) and (30)and (48) we obtained the equation for calculating the factor f in the following form:

$$\frac{1}{f^{0.5}} = \frac{Re_c}{(1/\eta - 1)\sqrt{G/\pi}}$$

(49)

To demonstrate how the friction factor reduces with an increase in the Reynolds number, we performed the two sets of Couette flow calculations for the two different radius ratios $\eta=0.5$ and 0.75. The computations were carried out for the same range of Reynolds numbers and for the same parameters α_{*_0} (except the highest $\alpha_{*_0}=17\times10^{-4}$), which were employed for illustration of drag reduction in a pipe flow (see Fig. 2). From Fig. 4 one can see that the obtained curves behave similarly to the corresponding graphs for a pipe flow. The thick straight line limiting the graphs from the top is the drag reduction asymptote indicating the maximum drag reduction effect achievable. This effect is well understood for pipe flows (e.g, Virk, 1971b). According to the estimations, given below, the drag reduction asymptote for a Couette device, formally, has the same form as the asymptote for a pipe flow that is described by the empirical equation ofVirk (1971b):

$$\frac{1}{f^{0.5}} = 19 \log_{10}(Re f^{0.5}) - 32.4$$

(50)

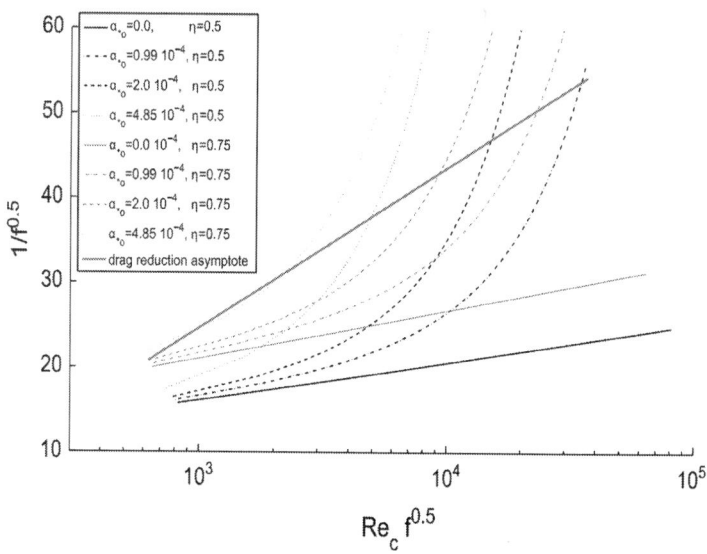

Figure 4: Characteristic lines of the drag reduction effect, caused by the different drag reducing chemicals, in a Couette device of the two different radius ratios.

Note that for a Couette device case, the Reynolds number for a Couette flow (Re_c) and the Fanning friction factor for the inner cylinder have to be employed for calculations. Justification of applicability of Eq. (50) to a Couette flow will be given further.

Thus, Eq. (50) provides the smallest friction factor obtainable by drag reducers. As was already explained in the Section 1, the maximum drag reduction is reached when the upper boundary of the buffer layer extends till the channel center, i.e., the turbulent boundary layer disappears. Although the physics of this phenomenon is clear, the modeling approach based on the boundary layer approximation, which we used in the current work, does not allow forecasting the drag reduction asymptote; therefore, Eq. (50) can be used to exclude from analysis unphysical results obtained by the drag reduction model.

Let us now derive the relation between the drag reduction asymptotes for Couette and pipe flows. The drag reduction effect stronger reveals itself at the inner Couette device wall, where the drag reduction factor is larger, than that at the outer wall (see Eqs. (41) and (42)). Then, it is possible to assume that the maximum drag reduction in a Couette flow is reached when the dimensionless viscous sublayer thickness δ_i^+ at the Couette device inner cylinder (see Eq. (43)) is equal to the viscous sublayer thickness δ_{max}^+ in a pipe flow corresponding to the maximum drag reduction regime.

The thickness δ_{max}^+ can be calculated by Eq. (12). The equality of δ_i^+ and δ_{max}^+ is reduced to equality of the maximum drag reduction factors achievable:

$$(D_{i*})^{max} = (D_*)^{max} \tag{51}$$

Using Eqs. (9) and (42) we obtain the following equation:

$$1 + \frac{\alpha_* (1 - \eta)}{2} \frac{}{\eta} \sqrt{\frac{G}{2\pi}} = 1 + \alpha_* R^+ \tag{52}$$

where $(1/2)(1-\eta)/\eta\sqrt{G/2\pi} = H^+/2$ is the dimensionless half-width of the Couette device gap.

Simplifying the left-hand side of Eq. (52) by Eq. (49) and rewriting the right hand-side of the same equation by using the well-known relation $R^+ = 0.5Re\sqrt{f/2}$, we obtain the simple equation:

$$(Re_c \sqrt{f})^{max} = (Re \sqrt{f})_{pipe}^{max} \tag{53}$$

Eq. (52) shows validity of the drag reduction asymptote, expressed by Eq. (50) for a Couette flow. Eq.(50) allows a significant improvement of the drag reducer testing technique using a Couette device. Fig. 4 shows that the smaller the Couette device radius ratio ($\eta=0.5$ vs. $\eta =0.75$) is, the stronger the curves are shifted downward from the drag reduction asymptote. This shift provides wider ranges of both the coefficient α_{*o} and the Reynolds number R_{ec} for testing different drag reducing chemicals. For example, it follows from Fig. 4 that the drag reducer, characterized by the relatively high coefficient α_{*_0}=4.85×10^{-4}, cannot be tested in a Couette device with the radius ratio η=0.75. One can see that, in this case, even at relatively low Reynolds numbers (the minimum Re_c is 13,000 for the plots in Fig. 4) the drag reduction model provides an unrealistically low friction factor (lower than the drag reduction asymptote gives). Then, it is easy to conclude that an efficient testing of drag reducing chemicals requires using a Couette device with a relatively low radius ratio η.

Finally, we would like to emphasize that in the present work we developed drag reduction models for pipe and Couette flows, which can be useful tools for scaling the drag reduction effect from a Couette device to an industrial pipeline. Although, no experimental data suitable for validation of the model of drag reduction in a Couette flow are available yet, there are strong indirect proofs of validity of the modeling approach developed. Both the models are based on the same assumptions, and the employed turbulence model is proven to be valid (Schlichting and Gersten, 2000 and Eskin, 2010) for both the flow types. The drag reduction modeling has been mainly reduced to description of the viscous sublayer flow affected by a polymer additive. Because the viscous sublayer is relatively thin, the sublayer flow patterns for a pipe and a Couette device are nearly identical. Therefore, because validity of the drag reduction model for a pipe flow has been confirmed by the experimental data (see Fig. 2), we can be confident that the same modeling approach employed for a Taylor–Couette flow is also correct.

CONCLUSIONS

An engineering model of turbulent drag reduction in a pipe has been developed. The model employs a well-known two-layer representation of the boundary layer structure. An approach of Yang and Dou (2010) to modeling the drag reduction effect, as a phenomenon caused by a non-Newtonian rheology of a viscous sublayer flow, is employed. This approach utilizes a known idea of stress deficit in boundary layer of a polymer solution flow. The modified Prandtl–Karman equation for calculation of the friction factor in a pipe flow of a dilute polymer solution has been obtained. The derived equation contains the only empirical parameter that is a function of polymer type and concentration. The results computed by the model developed are in a good agreement with those calculated by the Yang and Dou (2010) model and verified against experimental data. An engineering model of a turbulent dilute polymer solution flow in a Couette device has also been developed. The same approach to modeling drag reduction as that in a pipe flow was employed. The model allows to compute the dimensionless torque applied to the Coutte device rotor as a function of the rotation speed for a given polymer type and concentration.

It was also shown that the empirical drag reduction asymptote, known for a pipe flow, can be used for a Couette flow when the Fanning friction factor is calculated for the inner Couette device cylinder using its circumferential velocity. Also, the calculations showed that the Couette device with a smaller inner to outer radius ratio η is preferable for drag reducer testing because allows operating in regimes, which are further from the drag reduction asymptote than in the case of higher η. Thus, we recommend using a small-scale laboratory Couette device experiments for identifying the empirical parameter that characterizes the drag reduction caused by a certain polymer additive. The identified model parameter can be used for forecasting turbulent drag reduction in industrial-scale pipeline flows.

ACKNOWLEDGMENT

The author is thankful to Prof. Anthony Pearson (Schlumberger Gould Research Center) for a constructive discussion of this work.

REFERENCES

1. Benzi, R., Ching, E.S.C., Horesh, N., Procaccia, I., 2004. Theory of concentration dependence in drag reduction by polymers and the maximum drag reduction assymptote. Phys. Rev. Lett. 92, 078302.

2. Eskin, D., 2010. An engineering model of a developed turbulent flow in a Couette device. Chem. Eng. Process. 49, 219–224.

3. Gyr, A., Tsinober, A., 1997. On the rheological nature of drag reduction phenomena. J. Non-Newton. Fluid Mech. 73, 153–162.

4. Kalashnikov, V.N., 1998. Dynamical similarity and dimensionless relations for turbulent drag reduction by polymer additives. J. Non-Newton. Fluid Mech. 75, 1209–1230.

5. Koeltzsch, K., Qi, Y., Brodkey, R.S., Zakin, J.L., 2003. Drag reduction using surfactants in a rotating cylinder geometry. Exp. Fluids 24, 515–530.

6. Lathrop, D.P., Fineberg, J., Swinney, H.L., 1992. Transition to shear-driven turbulence in Couette–Taylor flow. Phys. Rev. A 46, 6390–6405.

7. Lewis, G.S., Swinney, H.L., 1999. Velocity structure functions, scaling and transitions in high-Reynolds-number Couette–Taylor flow. Phys. Rev. E 59, 5457–5467.

8. Liu, N., Khomami, B., 2013. Polymer-induced drag enhancement in turbulent Taylor–Couette flows: direct numerical simulations and mechanistic insight. Phys. Rev. Lett. 111, 114501.

9. Luchik, T.S., Tiederman, W.G., 1988. Turbulent structure in low-concentration dragreducing channel flows. J. Fluid Mech. 190, 241–263.

10. L'vov, V.S., Pomyalov, A., Procaccia, I., Tiberkevich, V., 2004. Drag reduction by polymers in wall-bounded turbulence. Phys. Rev. Lett. 92, 244503.

11. Procaccia, I., L'vov, V.S., 2008. Colloquium: theory of drag reduction by polymers in wall-bounded turbulence. Rev. Mod. Phys. 80, 225–247.

12. Reischman, M., Tiederman, W.G., 1975. Laser-Doppler anemometer measurements in drag-reducing channel flows. J. Fluid Mech. 70 (Part 2), 369–392.

13. Rudd, M.J., 1972. Velocity measurements made with a laser Doppler meter on the turbulent pipe flow of a dilute polymer solution. J. Fluid Mech. 51, 673–685.

14. Schlichting, H., Gersten, K., 2000. Boundary-Layer Theory. Springer-Verlag, Berlin, Heidelberg, New York.

15. Tabor, M, de Gennes, P.M., 1986. A cascade theory of drag reduction. Europhys. Lett. 2, 519–522.

16. Toms, B.A., 1948. Proceedings of the 1st International Congress on Rheology. North Holland.

17. Virk, P.S., 1971a. Drag reduction in rough pipes. J. Fluid Mech. 45, 225–246.

18. Virk, P.S., 1971b. An elastic sublayer model for drag reduction by dilute solutions of linear macromolecules. J. Fluid. Mech. 45, 417–440.

19. Willmarth, W.W., Wei, T., Lee, C.O., 1987. Laser anemometer measurements of Reynolds stress in a turbulent channel flow with drag reducing polymer additives. Phys. Fluids 30, 933–935.

20. Yang, S.-Q., Dou, G., 2010. Turbulent drag reduction with polymer additive in rough pipes. J. Fluid Mech. 642, 279–294.

21. Yang, S.-Q., Dou, G., 2008. Modeling of viscoelastic turbulent flow in channel and pipe. Phys Fluids 20, 065105.

22. Yang, S.-Q., Dou, G., 2005. Drag reduction in a flat-plate boundary layer flow by polymer additives. Phys. Fluids 17, 065104.

23. White, C.M, Mungal, M.G., 2008. Mechanics and prediction of turbulent drag reduction with polymer additives. Annu. Rev. Fluid Mech. 40, 235–256.

Chapter 5

Modelling the Impact of Stream Impurities on Ductile Fractures in CO_2 Pipelines

Haroun Mahgerefteh, Solomon Brown, and Garfield Denton

Department of Chemical Engineering, University College London, London WC1E 7JE, UK

ABSTRACT

This paper describes the development, validation and application of a fully coupled dynamic boundary fracture model for predicting ductile fracture behaviour in CO_2 pipelines. The application of the model to an hypothetical but realistic CO_2 pipeline reveals the profound effects of the line temperature and the types of impurities present in the CO_2 stream on the pipeline's propensity to fracture propagation. It is found that the pure CO_2 and the post-combustion pipelines exhibit very similar and highly temperature dependent propensity to fracture propagation. An increase in the line temperature from 20 to 30 °C results in the transition from a relatively short to a long running propagating facture. The situation becomes progressively worse in moving from the pre-combustion to the oxy-fuel stream. In the latter case, long running ductile fractures are observed at all the temperatures under consideration. All of the above findings are successfully explained by examining the fluid depressurisation trajectories during fracture propagation relative to the phase equilibrium envelopes.

INTRODUCTION

The deployment of Carbon Capture and Storage (CCS) is the cornerstone of the drive to reduce CO_2 emissions at least for the next three decades. As part of the CCS chain, pressurised pipelines are widely recognised as the most practical and economical means of transporting the huge amounts of captured CO_2 from coal fired power plants for subsequent sequestration (Barker et al., 2007 and Chandel et al., 2010). Typically, such pipelines may cover distances of several hundred kilometres at pressures above 100 bar.

Given that CO_2 is as an asphyxiant at high concentrations (Kruse and Tekiela, 1996), the safety assessment of CO_2 pipelines in the unlikely event of pipeline rupture is of paramount importance and indeed central to the public acceptability of CCS.

The above is particularly relevant given the likelihood of CO_2 pipelines passing through or near populated areas. Most electricity generation plants are built close to energy consumers. As such the number of people potentially exposed to risks from CO_2 transportation facilities will be greater than the corresponding number exposed to potential risks from CO_2 capture and storage facilities (Barker et al., 2007).

Ironically (in line with its abbreviation), CCS and related legislation generally focus on the capture and sequestration of CO_2 and not on its transportation.

It is noteworthy that CO_2 pipelines have been in operation in the US for over 30 years for enhanced oil recovery (Bilio et al., 2009 and Seevam et al., 2008). However, these are either confined to low populated areas, and/or operate below the proposed supercritical conditions (73.3 bar and 31.18 °C) that make CO_2 pipeline transportation economically viable, thus representing significantly less safety issues. Additionally, given their small number, it is not possible to draw a meaningful statistical representation of the risk. Parfomak and Fogler (2007) propose 'statistically, the number of incidents involving CO_2 should be similar to those for natural gas transmission'.

Given the urgency in meeting the CO_2 emission targets, a number of very recent CCS projects around the world have commenced that involve the use of pressurised pipelines for transporting captured CO_2 for sequestration. However so far most of these are either being considered (e.g. UK National Grid Scotland and Humberside CO_2 pipeline network (www.nationalgrid.com/corporate/About+Us/climate/CCS2/)), are under development (e.g. the Alberta Carbon Trunk Line in Alberta, Canada (www.enhanceenergy.com)) or involve the use of relatively short pipelines (e.g. the Total Lacq pilot project in south west France (http://www.total.com/en/special-reports/capture-and-geological-storage-of-co2/challenges-of-capture-and-geological-storage-of-co2-200960.html)).

At 150 km long and 22 cm i.d., the Snøhvit pipeline (de Koeijer et al., 2007 and et al., March 30, 2011) is the world's first long distance pipeline transporting dense phase CO_2 at 150 barg captured from a plant in Melkøya is land which is injected into a geological formation in the Barents Sea.

Running fractures are considered catastrophic pipeline failures. These involve the rapid tearing of the pipeline, sometimes running for several hundred metres resulting in the release of massive amounts of inventory in a very short space of time. Given its importance, a large number of studies spanning more than 30 years (see for example Leis et al., 2005 and Maxey, 1974) have been devoted to understanding the mechanism and overcoming such failures in the hydrocarbon industry.

In essence, such fractures can initiate from defects introduced into the pipe by outside forces such as mechanical damage, soil movement, corrosion, material defects or adverse operating conditions. When the stress acting on the defect overcomes the fracture toughness of the pipe the fracture will propagate, reaching a critical size based on the pipeline material properties and operating conditions. As such it is highly desirable to design pipelines such that when a defect reaches a critical size and fails, the result is a leak rather than a long running fracture.

In considering such failures, it is noteworthy that the temperature drop as a result of the Joule–Thomson expansion cooling of the fluid within the pipeline during discharge can be significant (Mahgerefteh and Atti, 2006). In the case of CO_2, depending on the starting conditions, such temperatures can reach as low as -70 °C resulting in very significant localised cooling of the pipe wall in contact with the escaping fluid.

The minimum pipe wall temperature reached relative to its ductile to brittle transition temperature will in turn dictate whether the pipeline will fail in the ductile or brittle fracture manner. In the case of ductile fractures, the 'global' pipe/wall heat transfer is important and has been accounted for in this work (Mahgerefteh et al., 2006). Given the rapid opening of the crack tip, there is simply insufficient time for the localised cooling of the crack tip by the discharging CO_2.

Of course the situation is somewhat different in the case of brittle fractures. Here accounting for localised heat transfer effects is of paramount importance when modelling such failures. The modelling of brittle fractures in pressurised pipelines has been presented in our previous publication (Mahgerefteh and Atti, 2006). Ductile fractures are the focus of attention in this work.

The so called Battelle Two Curve (BTC) method (Maxey, 1974) was the first used to express the criterion for the propagation of a ductile fracture in terms of the relation between the fluid decompression wave velocity and the crack propagation velocity. If the fluid decompression wave velocity is larger than the crack velocity, the crack tip stress will decrease, eventually dropping below the arrest stress causing the crack to arrest. Conversely, if the decompression wave velocity remains smaller than the crack velocity, the crack tip pressure will remain constant resulting in indefinite propagation.

Compared to natural gas for example, CO_2 has an unusually high saturation pressure. Depending on the starting pressure and temperature, the above coupled with the uniquely 'prolonged' depressurisation during the liquid/gas phase transition mean that CO_2 pipelines may be more susceptible to fracture propagation as compared to hydrocarbon pipelines (Cosham et al., 2010 and Mahgerefteh et al., 2010). As such accounting for any parameters that may modify the CO_2 depressurisation trajectory is of paramount importance when modelling ductile fractures in CO_2 pipelines.

One such important factor which is receiving increasing attention (see for example de Visser et al., 2008 and Heggum et al., 2005) is the impact of impurities. Recent studies using various equations of state have shown that even small amounts of the likely impurities in the CO_2 stream will increase the saturation pressure significantly (see for example Li and Yan, 2006 and Oosterkamp and Ramsen, 2008).

The type of these impurities will depend on the fuel, capture method (i.e. pre-combustion, post-combustion or oxy-fuel) and the post-capture processing (ICF International, 2010 and Oosterkamp and Ramsen, 2008). The percentage composition of the transported stream on the other hand, although being overwhelmingly CO_2, will have to comply with the prevailing legislative emission limits (de Visser et al., 2008 and ICF International, 2010).

So far most of the studies reported investigate the impact of impurities on the CO_2 phase equilibrium behaviour using various equations of state (Brown and Mahgerefteh, 2009; Li and Yan, 2009) and the resulting CO_2 decompression behaviour (Cosham and Eiber, 2008). In the latter case, a comparison of the resulting decompression and crack propagation velocity curves against pressure based on the Battelle Two Curve methodology (Maxey, 1974) is in turn used to infer the pipe toughness that would be required in order to arrest fracture (King and Kumar, 2010).

However such studies, although useful, employ over-simplistic decompression models where the impact of pipe wall heat transfer and friction on the fluid decompression behaviour is ignored (see for example King and Kumar, 2010, Makino et al., 2001a, Makino et al., 2001b and Terenzi, 2005). Based on shock tube experiments using different surface roughness pipes, Botros et al. (2010) for example have recently shown that ignoring pipe friction may result in overestimating the decompression wave velocity with the effect becoming more significant with increasing line pressure and reducing pipe diameter. This is important since an over-prediction of decompression wave velocity results in underestimating the pipe toughness that would be required in order to arrest fracture.

Crucially, given that the decompression and the fracture velocity curves are not coupled means that the Battelle Two Curve method is primarily indicative as to whether or not a fracture will propagate for a given pipe toughness. Important information such as the variation of the crack length with crack propagation velocity and, ultimately the crack arrest length cannot be produced using this approach.

In this paper we report the development and validation of a fully coupled Dynamic Boundary ductile Fracture Model (DBFM) taking into account heat transfer and friction. The model is next used to investigate the impact of various impurities on the fracture propagation and arrest in dense phase CO_2 pipelines.

It is appreciated that in some countries including the UK, for example, most of the CO_2 during on-shore transportation for off-shore sequestration will be in the gaseous phase. However, it is expected that in a future CCS infrastructure, the required flow rates will increase as a number of pipelines may tie into a single pipeline. As such, in order to cope with the increased capacity requirements, it is likely that further onshore compression of the

CO_2 to the dense phase will be required prior to transportation to the offshore sequestration site.

Clearly given the increased hazard associated with the dense phase, the safety of CO_2 pipelines during this stage of onshore transportation especially in the vicinity of populated areas will be of paramount importance.

The compositions of the impurities assumed are those based on the various capture technologies, including pre-combustion, post-combustion and oxy-fuel as suggested by ICF International (2010). In each case, the variation of fracture velocity versus fracture length is reported for an hypothetical CO_2 pipeline with a realistic fracture toughness. It is noteworthy that in practice, large diameter pipelines such as those envisaged for CO_2 transportation will be typically constructed from rolled steel in sections with a longitudinal weld. Each pipe section is then welded to the next via a circumferential weld with each longitudinal weld at 90° from its neighbour. The modelling of fracture propagation through a weld given its deferent metallurgical properties as compared to the rest of the pipe is highly complex and beyond the scope of this work.

Given the likely extensive variations in the geographical locations of the next generation CO_2 pipelines, the study also investigates the impact of the line temperature on the pipeline's propensity to fracture propagation.

THEORY

The Decompression Model

The full background theory of the fluid flow model employed in this study to predict the decompression behaviour, implemented in the CFD code PipeTech (http://www.pipetechsoftware.com/) including its validation against real pipeline rupture data is given elsewhere (Mahgerefteh et al., 2007, Mahgerefteh et al., 2006 and Oke et al., 2003). For completeness, a brief account of its main features is given here. Based on the homogeneous flow assumption, in the case of unsteady, one-dimensional flow the mass, momentum and energy conservation equations are respectively are given by

$$\frac{d\rho}{dt} + \rho\frac{\partial u}{\partial x} = 0$$

(1)

$$\rho\frac{\partial u}{\partial t} + \rho u\frac{\partial u}{\partial x} + \frac{\partial P}{\partial x} = \alpha$$

(2)

$$\rho\frac{dh}{dt} - \frac{dP}{dt} - (q_h - u\beta_y) = 0$$

(3)

where ρ, u, P and h are the density, velocity, pressure and specific enthalpy of the homogeneous fluid as function of time, t, and space, x. q_h is the heat transferred through the pipe wall to the fluid and β_y is the friction force term given by

$$\beta_y = -2\frac{f_w}{D}\rho u|u|$$

(4)

where, f_w is the Fanning friction factor and D the pipeline diameter.

Also

$$\alpha = -\left(\frac{2f_w\rho u|u|}{D} + \rho g\sin\theta\right)$$

where ϑ is the angle of inclination of the pipeline to the horizontal.

Eqs. (1), (2) and (3) are quasi-linear and must be solved numerically. In this study, the Method of Characteristics (MOC) (Zucrow and Hoffman, 1975) is used as the numerical solution method, as opposed to other numerical techniques such as finite element (Bisgaard and Sørensen, 1987 and Lang, 1991) and finite difference methods (Bendiksen et al., 1991, Chen et al., 1995a and Chen et al., 1995b) as both have difficulty in handling the choking condition at the rupture plane. The MOC handles the choked flow intrinsically via the Mach line characteristics. Moreover, MOC is considered to be more accurate than the finite difference method as it is based on the characteristics of wave propagation. Hence, numerical diffusion associated with a finite difference approximation of partial derivatives is reduced.

The Fracture Model

The key step in the BTC method is the derivation of two sets of curves; one set describing the crack velocity and the other the velocity of the fluid decompression wave. The resistance to crack propagation is indicated by the Charpy V-Notch (C_VN) energy (Maxey, 1974). However, in the full scale pipe bust tests conducted by the High-Strength Line Pipe Committee (HLP) (Makino et al., 2001a), the BTC theory is used in conjunction with the Drop Weight Tear Test (DWTT) energy, as this is shown to provide a more accurate indication of the pipeline resistance to fracture. Consequently this is the model applied in this work.

The two curve models for the crack propagation velocity, v_c and crack arrest pressure, P_a are respectively given by (Makino et al., 2001a, Makino et al., 2001b and Makino et al., 2001c)

$$v_c = 0.67 \frac{\sigma_{flow}}{\sqrt{D_p/A_P}} \left(\frac{P_t}{P_a} - 1\right)^{0.393}$$

(6)

$$P_a = 0.382 \frac{t_w}{D} \sigma_{flow} \cos^{-1} e^{((-3.81 \times 10^7 (D_p/A_p))/\sqrt{Dt}\sigma_{flow}^2)}$$

(7)

where σ_{flow}, D_p and A_p are respectively the flow stress (the mean value of the tensile and yield stresses), pre-cracked DWTT energy and ligament area of a pre-cracked DWTT specimen. On the other hand P_t and t_w are the crack tip pressure and pipe wall thickness respectively. Worse case scenario is assumed such that the initial defect in the pipe prior to growth is longitudinal, i.e. along the main axis.

The Coupled Decompression and Fracture Model

The calculation algorithm coupling the decompression and fracture models for simulating the fracture behaviour is given in Fig. 1. For simplicity, the crack opening is assumed to represent a full bore rupture which moves along the pipeline. This is a fair assumption given that any radial deflection or bending of the crack opening is not expected to play a role in the depressurisation process. The first step in the calculations involves the use of the MOC decompression model described above to calculate the rupture plane pressure, P_a and hence the crack tip pressure, P_t based on the given pipeline characteristics and the prevailing conditions. The crack velocity, v_c is then calculated by substituting P_t and P_a (in turn determined from Eq. (7)) into Eq. (6).

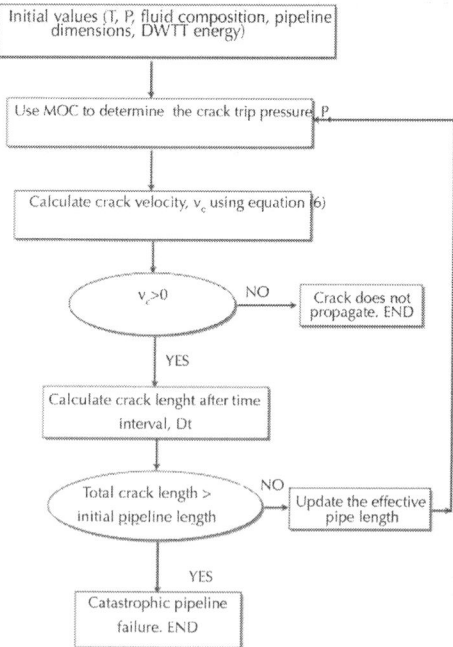

Figure 1: Crack propagation calculation flow algorithm.

A negative or zero crack velocity, v_c means that the crack will not extend and the calculations are terminated. A positive value on the other hand means that the crack will extend. Its length after an arbitrary small time interval, t (=0.01 s) during which v_c is assumed to remain constant in determined next. A calculated crack length greater than or equal to the pipe length means catastrophic pipeline failure and the calculations are ended. Otherwise, the new effective pipeline length is used to calculate the crack tip pressure, P_t. The calculations continue until $v_c \leq 0$.

RESULTS AND DISCUSSION

Validation

The following shows the results relating to the validation of the Dynamic Boundary ductile Fracture Model (DBFM) presented above by comparison of its predictions against the following published experimental data:

- HLP full scale burst test (Inoue et al., 2003).
- ECSC X100 pipe full scale burst test (Takeuchi et al., 2006).
- Alliance full scale burst tests (Johnson et al., 2000).

Table 1 shows the pertinent conditions relating to each test. Table 2 on the other hand shows the rich gas feed compositions for HLP C2 and Alliance tests.

Table 1: Pipeline characteristics and prevailing conditions utilised for the full scale burst tests

Parameter	HLP			ECSC	Alliance
	A1	B1	C2		
Inventory	Air	Air	Rich gas (see Table 2)	Air	Rich gas (see Table 2)

Internal diameter (m)	1.182	1.182	1.182	1.4223	0.8856
Pipe thickness (m)	0.0183	0.0183	0.0183	0.0191	0.0142
Initial pressure (bara)	116	116	104	126	120.2
Initial temperature (°C)	12	6	-5	20	23.9
Ambient pressure (bara)	1.01	1.01	1.01	1.01	1.01
Ambient temperature (°C)	20	20	20	20	20
Pipe length (m)	35	35	35	35	100
Tensile stress (MPa)	505	505	505	807	505
Yield stress (MPa)	482	482	482	728	482
Pipe grade	X70	X70	X70	X100	X70

Table 2: Rich gas feed compositions

Component	HLP C2	Alliance Test 1
CH_4	89.57	80.665
C_2H_6	4.7	15.409
C3H8	3.47	3.090
iC_4H_{10}	0.24	0.232
$nC4H10$	0.56	0.527
iC_5H_{12}	0.106	0.021
nC_5H_{12}	0.075	0.014
nC_6H_{14}	0.033	0.003

nC_7H16	0.017	0
nC_8H_{18}	0.008	0
nC_9H_{20}	0.001	0
N_2	0.5	0.039
CO_2	0.72	0

The full burst test pipelines used comprised several sections of differing toughness for which the corresponding DWTT energy may be calculated. In all simulations, the pipe wall roughness and heat transfer coefficient are taken as 0.05 mm and 5 W/ (m² K) respectively. The latter corresponds to the un-insulated pipeline being exposed to still air.

The HLP full scale experiments involved three series of burst tests, referred to as test series A, B and C using X70 API grade pipelines containing air and a rich gas mixture. Pipeline fracture was initiated using an explosive charge. Fig. 2 shows a schematic representation of the pipe setup.

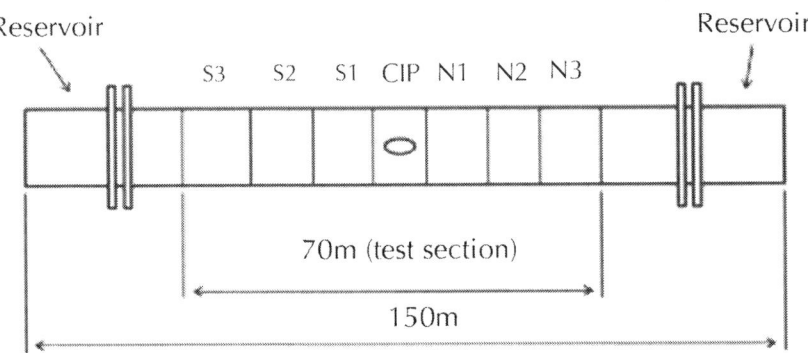

Figure 2: Schematic representation of the experimental setup used in the HLP full scale pipe burst tests (Inoue et al., 2003).

The Peng–Robinson Equation of State (Peng and Robinson, 1976) (PR EoS) is used for the prediction of the pertinent fluid phase equilibrium data for both air and rich gas mixtures. In the case of

CO_2, the Modified Peng–Robinson (Wu and Chen, 1997) EoS is used. As compared to the PR EoS, this equation has been shown to produce better predictions of the phase equilibrium data during the greater part of the depressurisation process (Mahgerefteh et al., 2008). An equi-distant grid system comprising 100 nodal points is employed for the fluid dynamic simulations. The corresponding discretisation time element is determined using 90% of the Courant, Friedrichs and Lewy value (Mahgerefteh et al., 2009).

Fig. 3, Fig. 4 and Fig. 5 show the variation of the crack velocity with crack length for the south running A1, B1 and C2 tests respectively for the HLP full scale experiments. Curves A show the measured crack length. Curves B on the other hand are the simulation predictions. In all cases, the corresponding Charpy Energy,C_V for each pipe section is given in the figures.

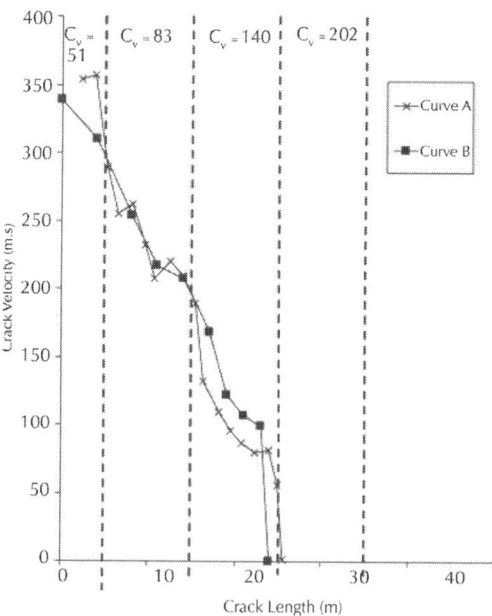

Figure 3: Variation of crack velocity with crack length for test A1 south running crack. Inventory: air, initial pressure=116 bara, initial temperature=12 °C. Curve A: experimental data (Inoue et al., 2003). Curve B: DBFM prediction.

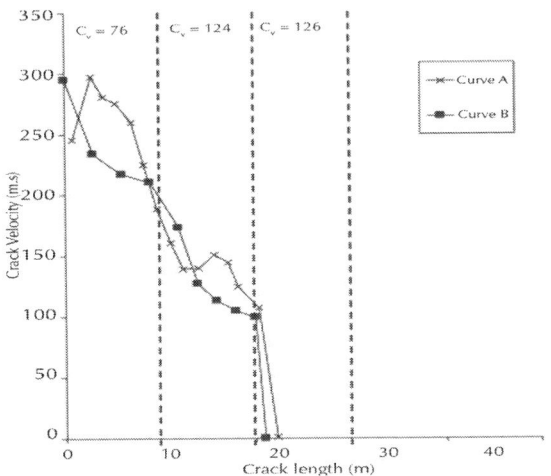

Figure 4: Variation of crack velocity with crack length for test B1 south running crack. Inventory: air, initial pressure=116 bara, initial temperature=6 °C. Curve A: experimental data (Inoue et al., 2003). Curve B: DBFM prediction.

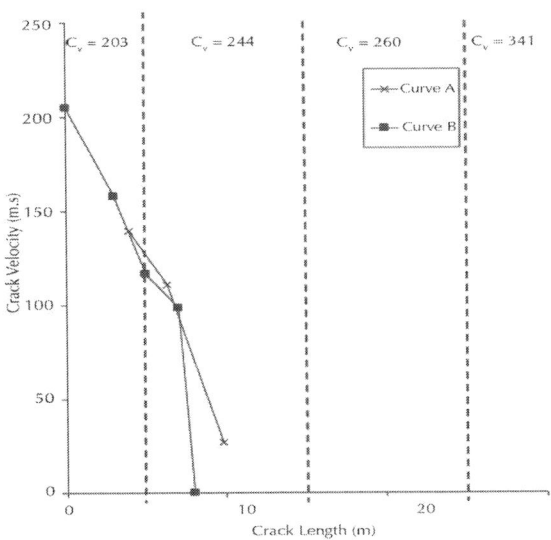

Figure 5: Variation of crack velocity with crack length for test C2 south running crack. Inventory: rich gas (Table 2), initial pres-

sure=104 bara, initial temperature=−5 °C. Curve A: experimental data (Inoue et al., 2003). Curve B: DBFM prediction.

Figure 6 and Fig. 7 show the corresponding data for ECSC X100 (Takeuchi et al., 2006) and Alliance full scale burst tests (Johnson et al., 2000) respectively.

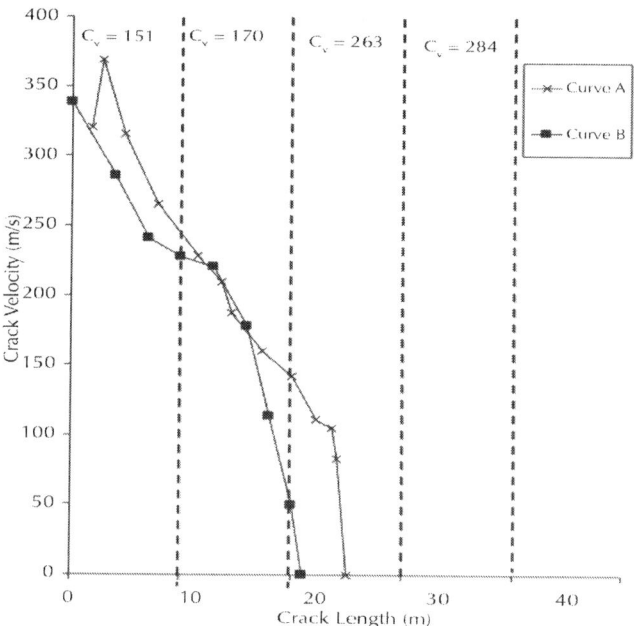

Figure 6: Variation of crack velocity with crack length for test ECSC X100 south running crack. Inventory: air, initial pressure=126 bara, initial temperature=20 °C. Curve A: experimental data (Makino et al., 2008). Curve B: DBFM prediction.

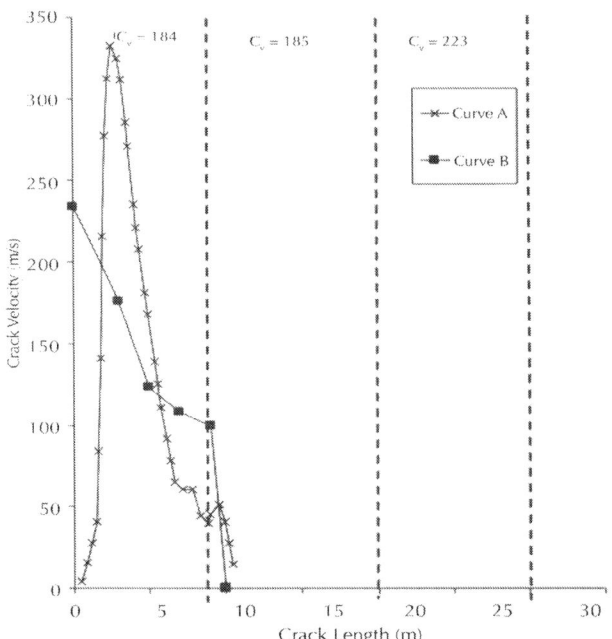

Figure 7: Variation of crack velocity with crack length for test Alliance Test 1. Inventory: rich gas (Table 2), initial pressure=120.2 bara, initial temperature=23.9 °C. Curve A: experimental data (Johnson et al., 2000). Curve B: DBFM prediction.

Returning to Fig. 3, Fig. 4, Fig. 5, Fig. 6 and Fig. 7, as it may be observed, the crack velocity significantly decreases with increase in crack length. This is due to the significant rapid decrease in the crack tip pressure as the pipeline depressurises. As an example, such behaviour expressed in terms of the variation of the cark tip pressure with time is shown in Fig. 8 for the HLP A1 south running crack.

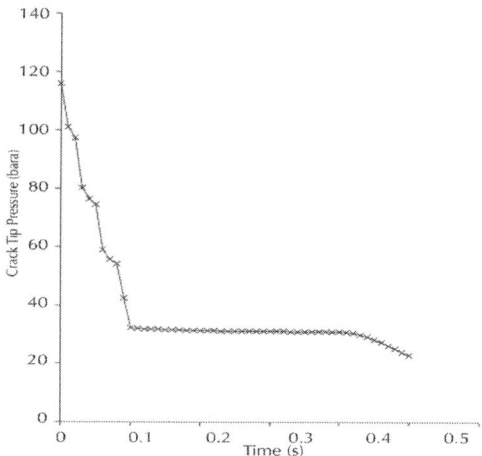

Figure 8: Simulated variation of crack tip pressure versus time for test HLP A1 south running crack

Also based on Fig. 3, Fig. 4, Fig. 5, Fig. 6 and Fig. 7 as expected, the crack velocity decreases as the crack propagates into the pipeline section with the higher toughness eventually coming to rest in all cases. As expected, the data in Fig. 5 show the smallest crack length as compared to the other tests due to the combination of the much higher fracture toughness pipe material employed coupled with the lowest initial pressure.

The initial rapid increase in the crack velocity observed in many of the test data is due to the finite time taken for the initial notch to fully develop into an open flap following detonation. This time domain is ignored in the present simulations.

Returning to the simulation data presented in Fig. 3, Fig. 4, Fig. 5, Fig. 6 and Fig. 7 (curves A), given the experimental uncertainties, it is clear that in all cases the DBFM predictions produce reasonably good agreement with the test data (curves B).

Impact of Impurities on Ductile Fractures

The following presents the simulation results obtained based on the application of the DBFM described above to a hypothetical pipeline containing various dense phase mixtures of CO_2 at 100 barg in the temperature range 0–30 °C.

The pipeline characteristics and the prevailing conditions are given in Table 3. The adopted pipeline grade, X65, the line pressure and the overall pipeline dimensions are those considered to be the most likely (Cosham and Eiber, 2008) to be employed for CO_2 pipelines as part of the CCS chain. However, in order to maintain practical computational run times, the pipeline length is limited to 500 m. The un-insulated pipeline is assumed to be exposed to still air corresponding to a heat transfer coefficient of 5 W/ (m² K). The ambient temperature is assumed to be the same as the line temperature.

Table 3: Pipeline characteristics and prevailing conditions utilised for the fracture propagation simulations

Parameter	Value
Internal diameter (m)	0.5905
Wall thickness (mm)	9.45
Line pressure (barg)	100
Ambient pressure (bara)	1.01
Ambient temperature (°C)	20
Feed temperature (°C)	0,10,20,30
Pipe length (m)	500
Tensile stress (MPa)	531
Yield stress (MPa)	448
Pipe wall roughness (mm)	0.05
Heat transfer coefficient (W/m²K)	5
Wind speed (m/s)	0
Pipe grade	X65
Fracture toughness (J)	50

Table 4 shows the various CO_2 stream compositions employed in the fracture simulations according to post-combustion, pre-combustion and oxy-fuel capture technologies (ICF International, 2010).

Table 4: CO_2 stream % compositions based on the various capture technologies (ICF International, 2010)

Species	Post-combustion	Pre-combustion	Oxy-fuel
CO_2	99.82	95.6	88.4
Ar	0	0	3.7
CO	0	0.4	0
N_2	0.17	0.6	2.8
H_2S	0	3.4	0
Cl	0	0	0.14
H_2	0	0	0
O_2	0.01	0	3.6
SO_2	0	0	1.36
H_2O	0	0	0
NO_2	0	0	0

An equi-distant grid system comprising 500 nodal points is employed for the fluid dynamic simulations. The corresponding discretisation time element required to ensure numerical stability is determined based on the Courant, Friedrichs and Lewy criterion (Mahgerefteh et al., 2009).

Fig. 9, Fig. 10, Fig. 11 and Fig. 12 respectively show the impact of the line temperature in the range 0–30 °C on the variation of the crack velocity with crack length for the pure CO_2, post-combustion, pre-combustion and oxy-fuel mixtures.

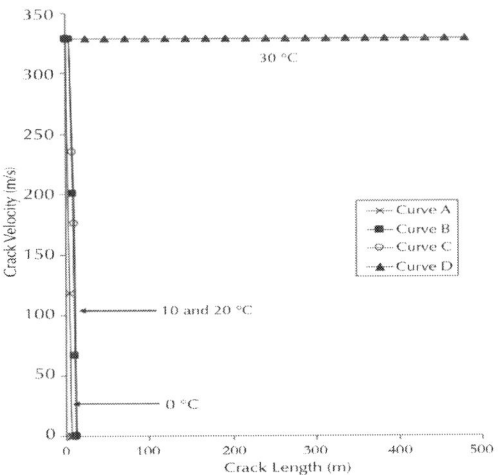

Figure 9: The impact of the line temperature on the variation of crack velocity with crack length for pure CO_2 pipeline at 100 barg. Curve A: 0 °C, Curve B: 10 °C, Curve C: 20 °C, Curve D: 30 °C.

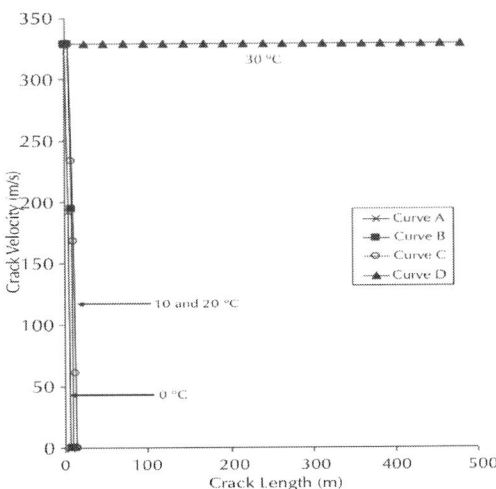

Figure 10: The impact of the line temperature on the variation of crack velocity with crack length for post-combustion CO_2 pipeline at 100 barg. Curve A: 0 °C, Curve B: 10 °C, Curve C: 20 °C, Curve D: 30 °C.

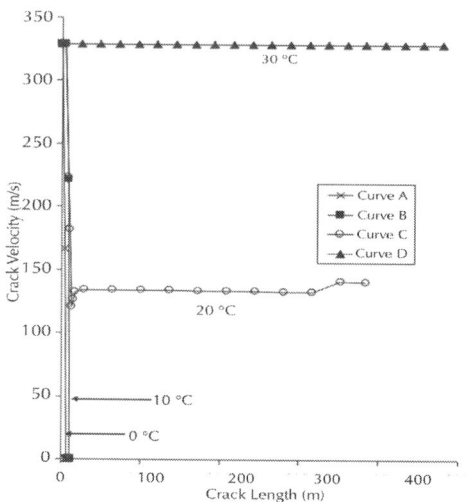

Figure 11: The impact of the line temperature on the variation of crack velocity with crack length for pre-combustion CO_2 pipeline at 100 barg. Curve A: 0 °C, Curve B: 10 °C, Curve C: 20 °C, Curve D: 30 °C.

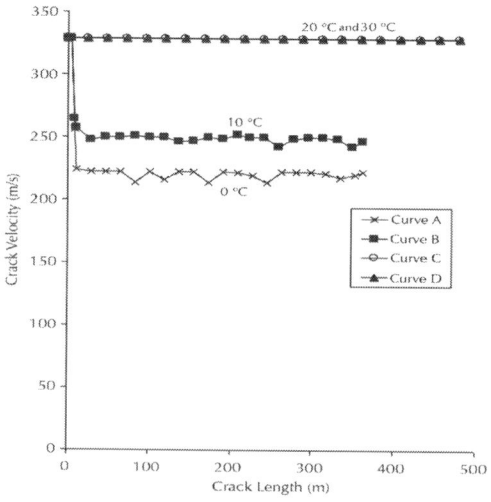

Figure 12: The impact of the line temperature on the variation of crack velocity with crack length for oxy-fuel combustion CO_2 pipe-

line at 100 barg. Curve A: 0 °C, Curve B: 10 °C, Curve C: 20 °C, Curve D: 30 °C.

The following pertinent observations may be made based on the data presented in the figures:

- Referring to Fig. 9 for pure CO_2, in the temperature range 0–20 °C (curves A–C), an increase in the line temperature results in a modest increase in the crack arrest length and fracture velocity. The maximum crack length is limited to a distance of 13 m. However, remarkably only a 10 °C rise in the line temperature to 30 °C (curve D) results in a fast running fracture travelling through the entire length of the 500 m pipeline.

- Based on the data presented in Fig. 10 for the post-combustion CO_2 stream, the fracture behaviour is substantially the same as that for the pure CO_2 (Fig. 9) with the expectation of the crack length running over a marginally longer distance between 0 and 20 °C before coming to rest. Once again the rise in the line temperature to 30 °C results in fracture propagating through the entire pipeline length.

- In the case of the pre-combustion CO_2 stream; Fig. 11, the transition to a long running facture commences at a lower temperature of 20 °C as compared to the pure (Fig. 9) and post-combustion (Fig. 10) CO_2 streams.

- According to Fig. 12, the oxy-fuel composition demonstrates the worst case scenario. Here long running fractures are obtained at all the temperatures tested.

- Table 5 shows a summary of the data presented in Fig. 9, Fig. 10, Fig. 11 and Fig. 12 expressed in terms of the variation of ratio of crack length to pipeline length for the various capture technologies. A ratio of unity means a crack propagating through the entire pipeline length.

Table 5: Ratio of crack to pipeline length for pure CO_2 and the various capture technologies in the temperature range 0–30 °C

Capture technology	Temperature (°C)	Ratio of crack to pipeline length
100% CO2	0	0.012
	10	0.024
	20	0.026
	30	1
Post-combustion	0	0.0012
	10	0.02
	20	0.028
	30	1
Pre-combustion	0	0.012
	10	0.022
	20	1
	30	1
Oxy-fuel	0	1
	10	1
	20	1
	30	1

To explain the above behaviour, Fig. 13 shows the fluid depressurisation trajectories expressed in terms of the variation of crack tip pressure with the fluid crack tip temperature at different starting line temperatures of 0 °C (curve A), 10 °C (curve B), 20 °C (curve C) and 30 °C (curve D) for the CO_2 pipeline. Curve E shows the vapour/liquid saturation curve. The crack arrest pressure of 43.65 barg calculated from Eqs. (6) and (7)is also shown for comparison. The corresponding data for the pre-combustion, post-combustion and oxy-fuel mixtures is given in Fig. 14, Fig. 15 and Fig. 16 respectively. The appearance and the subsequent broadening of the phase transition envelop in moving from the pre-combustion to the oxy-fuel compositions is due to the presence of increasing amounts of impurities in the CO_2 stream.

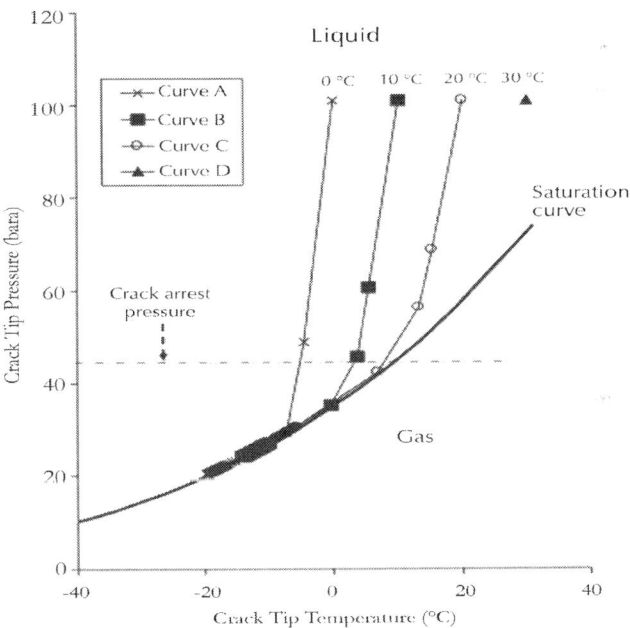

Figure 13: The variation of crack tip pressure with temperature for the 500 m long, 100 barg CO_2 pipeline at different starting line temperatures. Curve A: 0 °C, Curve B: 10 °C, Curve C: 20 °C, Curve D: 30 °C.

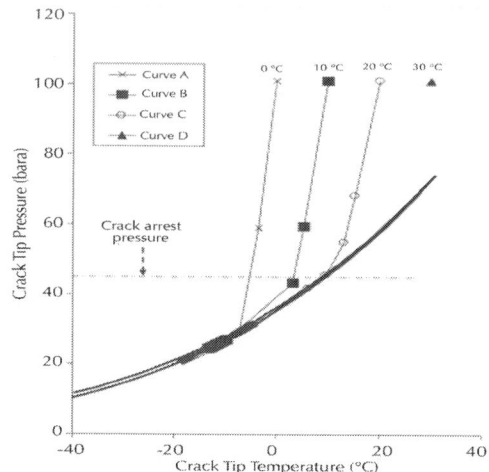

Figure 14: The variation of crack tip pressure with temperature for the 500 m long, 100 barg post-combustion pipeline at different starting line temperatures. Curve A: 0 °C, Curve B: 10 °C, Curve C: 20 °C, Curve D: 30 °C.

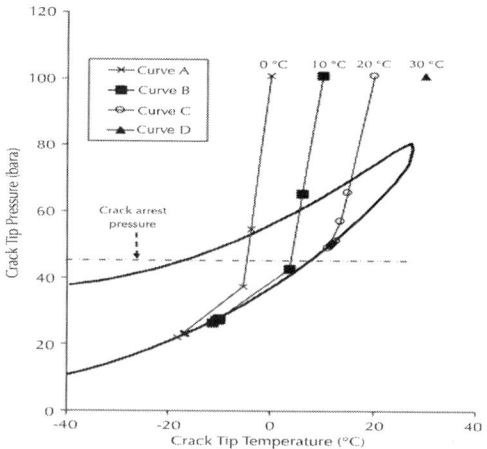

Figure 15: The variation of crack tip pressure with temperature for the 500 m long, 100 barg pre-combustion pipeline at different starting line temperatures. Curve A: 0 °C, Curve B: 10 °C, Curve C: 20 °C, Curve D: 30 °C.

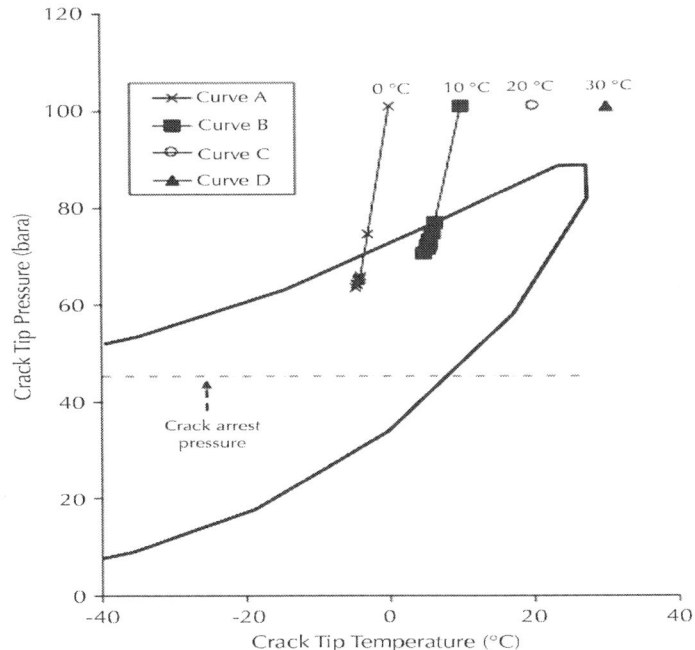

Figure 16: The variation of crack tip pressure with temperature for the 500 m long, 100 barg oxyfuel pipeline at different starting line temperatures. Curve A: 0 °C, Curve B: 10 °C, Curve C: 20 °C, Curve D: 30 °C

Returning to Fig. 13 as it may be observed, for pure CO_2 in the 0–20 °C range, the onset of fracture results in a rapid drop in the crack tip pressure from the dense phase to the saturated state. The accompanying significant drop in the fluid temperature is due to its near adiabatic expansion. The above is followed by the relatively slow decompression along the saturation curve where the liquid/vapour transition takes place. Based on the data presented, it is important to note that the pressure at which the depressurisation trajectories intersect the saturation curve increases with increase in the starting line temperature. Given that the crack will propagate until such time that the crack tip pressure is equal to the crack arrest pressure, it is to be expected that the crack length must increase with increase in the line temperature. This is consistent with the data shown in Fig. 9 presenting the variation of the crack velocity versus crack length. In the case of the pure CO_2 stream at 30 °C (Fig. 5, curve D), the crack tip pressure remains well above the crack arrest pressure of 43.65 barg throughout the depressurisation. Hence the crack will propagate through the entire pipeline length as indicated in Fig. 9 (curve D).

Exactly the same arguments apply in explaining the trends observed in Figs. 14–16. Given the relatively low level of impurities, the post-combustion pipeline (Fig. 14) exhibits a very similar propensity to fracture propagation to the CO_2 pipeline (Fig. 13) where propagating crack occurs only when the line temperature reaches 30 °C.

CONCLUSIONS

The safety of the next generation CO_2 pipelines is key to the success of CCS as a viable means for combating the effects of global warming. Put simply, given the heightened public concern towards the introduction of such new technology, even a single incident involving a major release of CO_2 near a populated area could jeopardise the large scale introduction of this nascent technology.

Of all the possible modes of pipeline failure, long running ductile fractures are catastrophic. As such the development of reliable quantitative techniques for predicting such failures for CO_2 pipelines in the presence of the typical likely impurities is of significant value.

Most of the investigations reported so far on this subject are based on the long-established Battelle Two Curve approach. This involves a comparison of the fluid decompression wave and fracture velocities vs. pressure curves as a function of the pipe fracture toughness. Given that the decompression and the velocity curves are not coupled this methodology is only capable of predicting the spontaneity of a given pipeline to fracture propagation.

In this paper we reported the development and testing of a fully coupled validated dynamic boundary ductile fracture propagation model which accounts for all the important fluid/structure interactions taking place during fracture propagation. These include, unsteady real fluid flow, heat transfer, friction as well as the drop in the crack tip pressure loading as the effective pipeline length reduces during fracture propagation.

Based on the application of the model to an hypothetical pipeline with a realistic fracture toughness transporting dense phase pure CO_2 as well as those based on pre-combustion, post-combustion and oxy-fuel capture technologies, the following general conclusions may be made:

- In the case of pure CO_2, the starting line temperature has a profound effect on the fracture behaviour. Whereas in the temperature range 0–20 °C the fracture length is limited to very short distances, only a 10 °C rise in the temperature from 20 to 30 °C results in a fracture propagating through the entire pipeline length. Given the relatively small amounts of impurities present, the post-combustion pipeline behaves in much the same way. Clearly, the above finding has important implications for pipelines routes where the ambient temperatures may exceed 20 °C.

- The transition to a long running fracture in the case of pre-combustion CO_2 is lower than that for the pure and post-combustion CO_2. In this case, long running fractures is likely at any temperature above 10 °C.

- The oxy-fuel composition represents the worst case scenario. Here long running fractures would be expected under all the temperatures under consideration.

- It is important to point out that the above findings are based on the particular conditions simulated in this study. Pipeline material of construction and compression costs will play important roles in dictating the precise CO_2 composition that may be safely and economically transported. Furthermore, there will be safety and environmental restrictions governing the compatibility of the CO_2 impurities with those that may be securely contained in the storage site. The above mean that it is likely that the precise CO_2 composition ultimately transported in the pipeline may be different from that produced by the power plant.

Given the above, the findings in this paper should not be used as the basis for selecting fracture toughness for CO_2 pipelines. This is especially true given the marked and unpredictable impact of the line temperature and composition on the CO_2 depressurisation thermodynamic trajectory and the resulting pipeline's propensity to fracture propagation. Any change in the pipeline operating pressure, not investigated in this study, is also an additional parameter which can and will have a significant and possibly unexpected impact.

As such, in practice all the relevant CO_2 pipeline design and operating parameters must be taken into account when selecting pipeline materials with the required toughness to withstand long running ductile fractures. Reliable predictive models such as the one presented in this work may serve as powerful tools for providing such information in a quantitative and inexpensive manner. Certainly the observed significant impact of the line temperature will have serious implications on the safety of CO_2 pipelines in locations where the ambient temperature is likely to rise above 20 °C.

Finally, the failure scenario considered in this study is that of ductile fracture propagation in a long distance pipeline during normal operation in which the pipeline inventory is at the same temperature as the pipe wall and the surrounding ambient. Clearly this assumption will not apply near compressor stations where significant localised heating of the CO_2 stream may occur. Also large differences in the ambient temperature are expected during the crossing of the pipeline from the warm onshore to the relatively cold sub-sea environment. Such temperature changes may cause two-phase flow.

The impact of fluid/ambient temperature differences and temperature gradients within the flowing fluid in CO_2 pipelines on ductile fracture propagation behaviour is the subject of a current study by the authors.

ACKNOWLEDGMENT

The authors are grateful to the EC FP7 framework programme (Grant no. 241346-2-CO2PipeHaz) for the provision of the funding required to carry out this work.

REFERENCES

1. Barker, T., Bashmakov, I., Bernstein, L., Bogner, J., Bosch, P., Dave, R., et al., 2007. Summary for Policymakers Drafting, Contribution of Working Group III to the Fourth Assessment Report of the Intergovernmental Panel on Climate Change. Energy, Bangkok.

2. Bendiksen, K., Maines, D., Moe, R., 1991. The dynamic two-fluid model OLGA: theory and application. SPE Prod. 6 (6), 171–180. (Society of Petroleum Engineers).

3. Bilio, M., Brown, S., Fairwheather, M., Mahgerefteh, H., 2009. CO2 pipelines material and safety considerations. IChemE Symposium Series: HAZARDS XXI. Process Saf. Environ. Prot. 155, 423–429. (Manchester, IChemE).

4. Bisgaard, C., Sørensen, H., 1987. A finite element method for transient compressible flow in pipelines. Int. J. Numer. Methods Fluids 7 (1986), 291–303.

5. Botros, K.K., Geerligs, J., Rothwell, B., Carlson, Lorne, Fletcher, L., Venton, P., 2010. Transferability of decompression wave speed measured by a small-diameter shock tube to full size pipelines and implications for determining required fracture propagation resistance. Int. J. Pressure Vessels Piping 87 (12), 681–695.

6. Brown, S., Mahgerefteh, H., 2009. From cradle to burial high pressure equilibrium behaviour of CO2 during CCS. In: Proceedings of the 2009 AIChE Annual Meeting, Nashville.

7. Chandel, M.K., Pratson, L.F., Williams, E., 2010. Potential economies of scale in CO2 transport through use of a trunk pipeline. Energy Convers. Manage. 51 (12), 2825–2834.

8.   Chen, J.R., Richardson, S.M., Saville, G., 1995a. Modelling of two-phase blowdown from pipelines—I. A simplified numerical method for method for multicomponent mixtures. Chem. Eng. Sci. 50 (13), 2173–2187.

9.   Chen, J., Richardson, S., Saville, G., 1995b. Modelling of two-phase blowdown from pipelines—I. A hyperbolic model based on variational principles. Chem. Eng. Sci. 50 (4), 695–713.

10. Cosham, A., Eiber, R., 2008. Fracture propagation in CO2 pipelines. J. Pipeline Eng. 7, 115–124.

11. Cosham, A., Eiber, R.J., Clark, E.B., 2010. GASDECOM: Carbon dioxide and other components. In: Proceedings of the 8th International Pipeline Conference. Calgary, pp. 1–18.

12. De Visser, E., Hendriks, C., Barrio, M., Molnvik, M., Dekoeijer, G., Liljemark, S., et al., 2008. Dynamis CO2 quality recommendations. Int. J. Greenhouse Gas Control 2 (4), 478–484.

13. De Koeijer, G., Borch, J.H., Jakobsenb, J., Hafner, A., 2007. CO2 pipeline test rig for R&D and operator training. Transmission of CO2, H2 and biogas: exploring new uses for natural gas pipelines. In: Global Pipeline Monthly and Clarion Technical Conferences. Amsterdam.

14. Heggum, G., Weydahl, T., Roald, W., Mølnvik, M., Austegard, A., 2005. CO2 conditioning and transportation. In: Thomas, D.C., Benson, S.M. (Eds.), Carbon Dioxide Capture for Storage in Deep Geologic Formations, vol. 2; 2005. /http://www.total.com/en/special-reports/capture-and-geological-storage-of-co2/ challenges-of-capture-and-geological-storage-of-co2-200960.htmlS. Retrieved March 30, 2011.

15. Inoue, T., Makino, H., Endo, S., Kubo, T., Matsumoto, T., 2003. Simulation method for shear fracture propagation in natural gas transmission pipelines. Int. Offshore Polar Eng. Conf. 5, 121–128. (Honolulu).

16. ICF International, 2010. Implementation of Directive 2009/31/EC on the Geological Storage of Carbon Dioxide.

Guidance Document 2. Site Characterisation, CO2 Stream Composition, Monitoring and Corrective Measures.

17. Johnson, D.M., Horner, N., Carlson, L., Eiber, R., 2000. Full scale validation of the fracture control of a pipeline designed to transport rich natural gas. Pipeline Technol. 1, 331.

18. King, G., Kumar, S., 2010. How to select wall thickness, steel toughness, and operating pressure for long CO2 pipelines. J. Pipeline Eng. 9 (4), 253–264.

19. Kruse, H., Tekiela, M., 1996. Calculating the consequences of a CO2-pipeline rupture. Energy Convers. Manage. 37 (95), 1013–1018.

20. Lang, E., 1991. Gas flow in pipelines following a rupture computed by a spectral method. J. Appl. Math. Phys 42. (March).

21. Leis, B., Zhu, X., Forte, T., 2005. Modeling Running Fracture in Pipelines–Past, Present, and Plausible Future Directions. icf11.com.

22. Li, H., Yan, J., 2006. Impact of impurities in CO2-fluids on CO2 transport process. In: Proceedings of GT2006, Barcelona, pp. 367–375. H. Mahgercfteh et al. / Chemical Engineering Science 74 (2012) 200–210 209

23. Li, H., Yan, J., 2009. Evaluating cubic equations of state for calculation of vapor– liquid equilibrium of CO2 and CO2-mixtures for CO2 capture and storage processes. Appl. Energy 86 (6), 826–836.

24. Mahgerefteh, H., Atti, O., Denton, G., 2007. An interpolation technique for rapid CFD simulation of turbulent two-phase flows. Process Saf. Environ. Prot. 85 (1), 45–50.

25. Mahgerefteh, H., Denton, G., Rykov, Y., 2008. CO2 pipeline rupture. IChemE Symposium Series: HAZARDS XX. Process Saf. Environ. Prot., 869–879. (Manchester, IChemE).

26. Mahgerefteh, H., Oke, A., Rykov, Y., 2006. Efficient numerical solution for highly transient flows. Chem. Eng. Sci. 61 (15), 5049–5056.

27. Mahgerefteh, H., Brown, S., Zhang, P., 2010. A dynamic boundary ductile-fracturepropagation model for CO2 pipelines. J. Pipeline Eng. 9 (4), 265–276.

28. Mahgerefteh, H., Atti, O., 2006. Modeling low-temperature-induced failure of pressurized pipelines. AIChE J. 52 (3), 1248–1256.

29. Mahgerefteh, H., Rykov, Y., Denton, G., 2009. Courant, Friedrichs and Lewy (CFL) impact on numerical convergence of highly transient flows. Chem. Eng. Sci. 64 (23), 4969–4975.

30. Makino, H., Kubo, T., Shiwaku, T., Endo, S., Inoue, T., Kawaguchi, Y., et al., 2001a. Prediction for crack propagation and arrest of shear fracture in ultra-high pressure natural gas pipelines. ISIJ Int. 41 (4), 381–388.

31. Makino, H., Sugie, T., Watanabe, H., Kubo, T., Shiwaku, T., Endo, S., et al., 2001b. Natural gas decompression behavior in high pressure pipelines. ISIJ Int. 41 (4), 389–395.

32. Makino, H., Takeuchi, I., Tsukamoto, M., Kawaguchi, Y., 2001c. Study on the propagating shear fracture in high strength line pipes by partial-gas burst test. ISIJ Int. 41 (7), 788–794.

33. Makino, H., Takeuchi, I., Higuchi, R., 2008. Fracture propagation and arrest in highpressure gas transmission pipeline by ultra-strength line pipes. In: 7th International Pipeline Conference, Calgary.

34. Maxey, W.A., 1974. Fracture initiation, propagation and arrest. In: Proceedings of the 5th symposium in Line Pressure Research. Houston.

35. Oke, A., Mahgerefteh, H., Economou, I., Rykov, Y., 2003. A transient outflow model for pipeline puncture. Chem. Eng. Sci. 58, 4591–4694.

36. Oosterkamp, A., Ramsen, J., 2008. State-of-the-art overview of CO2 pipeline transport with relevance to offshore pipelines. Security.

37. Parfomak, P.W., Fogler, P., 2007. Carbon Dioxide (CO2) Pipelines for Carbon Sequestration: Emerging Policy Issues. CRS Report for Congress. Order Code RL33971.

38. Peng, D.Y., Robinson, D.B., 1976. A new two-constant equation of state. Industrial and Engineering Chemical Fundamentals 15, 59.

39. Seevam, P.N., Race, J.M., Downie, M.J., Hopkins, P., 2008. Transporting the next generation of CO2 for carbon, capture and storage: the impact of impurities on supercritical CO2 pipelines. ASME Conf. Proc. 2008 (48579), 39–51. (ASME).

40. Takeuchi, I., Makino, H., Okaguchi, S., Takahashi, N., Yamamoto, A., 2006. Crack arrestability of high-pressure gas pipelines by X100 or X120. 23rd World Gas Conference. Amsterdam. Terenzi, A., 2005. Influence of real-fluid properties in modeling decompression wave interacting with ductile fracture propagation. Oil Gas Sci. Technol. 60 (4), 711–719.

41. Wu, D., Chen, S., 1997. A modified Peng–Robinson equation of state. Chem. Eng. Commun. 156 (1), 215–225. /www.enhanceenergy.comS. Retrieved March 30, 2011. / www.nationalgrid.com/corporate/AboutþUs/climate/ CCS2/S. Retrieved March 30, 2011. /www.statoil.com/en/ ouroperations/explorationprod/ncs/snoehvit/pages/default. aspxS. Retrieved March 30, 2011.

42. Zucrow, M.J., Hoffman, J.D., 1975. Gas Dynamics. Wiley, New York.

Chapter 6

Transient Analysis of Isothermal Gas Flow in Pipeline Network

S.L Ke and H.C Ti

Department of Chemical and Environmental Engineering, National University of Singapore, Singapore

ABSTRACT

Conventional methods for transient analysis of pipeline network are normally applied to find the numerical solution of the two partial differential equations (continuity and momentum), which are complex and cumbersome. Following the success of the steady state analysis of pipeline network, this study extends the usage of the electrical analogy method by combining resistance with the theoretically derived models of capacitance and inductance. This method leads to a set of first-order ordinary differential equations for transient analysis of isothermal gas flows in pipeline network. Solving the proposed first-order ordinary differential equation is definitely much simpler than solving the set of partial differential equations. The computational advantages of the present method are demonstrated by comparing them with the conventional methods when applied to a range of pipe network simulation examples.

INTRODUCTION

The analysis of flows and pressure drops in piping systems has been studied by many workers and is usually based upon the consideration of steady state conditions. However, the steady state analysis of pipeline network is less applicable, in the design of actual transmission systems, as unsteady (transient) state is more often encountered.

The analysis of unsteady state is much more difficult than that of steady state. The reason for this difficulty is that the unsteady state system is typified by variables which are functions of time and space (or position). In contrast, for steady state analysis, the variable in the system is only a function of space. The introduction of the concept of time variable adds on a new dimension to the mathematical model of the transient flow in a pipe distribution system, and results in computational difficulty. The mathematical model of pipe distribution system is derived from the two conservation equations (mass and momentum), which have the following forms [8]:

$$\frac{\partial \rho}{\partial t} + \frac{\partial (\rho v)}{\partial x} = 0 \tag{1a}$$

$$\frac{\partial (\rho v)}{\partial t} + \frac{\partial (\rho v^2)}{\partial x} + \frac{\partial P}{\partial x} + 2\rho v^2 \left[\frac{\varphi_f}{D} \right] + \rho g \sin\theta = 0 \tag{1b}$$

In which ρ is the fluid density, v(x, t) the fluid velocity, P(x, t) is the pressure, $_f$ is Fanning friction factor, D is the pipe diameter, g is the gravitational acceleration constant, and ϑ is the angle between the horizon and the longitudinal direction of the pipe. Many algorithms such as finite difference methods or method of characteristic (MOC) have been used to solve the above partial differential equations [1], [8] and [13]. These methods have been shown to be efficient in solving unsteady flow equations. However, it is not the objective of the present study to follow the conventional route for the analysis of transient flow in pipe distribution system, but to apply electrical analogies to simulate the same transient flow problem.

ELECTRICAL ANALOGY

The analogy between fluid network and electrical network has been successfully applied in the simulation of steady state pipeline networks by many workers [2], [3] and [11]. From the electrical circuit theory, the three basic elements which relate to voltage and current are viz. resistance, capacitance and inductance. Based on the analogy of voltage and current in an electrical circuit network with that of pressure drop and flow in the fluid network, all the three basic elements should also be present in the fluid distribution systems [5], [6], [7] and [10].

The resistance effect of a pipeline, which has been studied in the steady state analysis of pipeline network, is due to several factors, such as the roughness and geometric properties of a pipe, the viscosity of the fluid and the flow rate. The capacitance effect of a pipeline can be directly attributed to the compressibility of the fluid. The inductance effect of a pipeline is believed to be due to the kinetic energy of the fluid [12].

DERIVATION OF MODELS FOR BASIC ELEMENTS

The models of resistance and capacitance have been introduced in a previous work [10]. However, it is intended that the models of the three basic elements be derived from and directly, such that the formulation of the models is explicated theoretically.

Due to pressure change involved in a transient process, the control volume may be compressed or expanded, so the Continuity equation, that is Eq. (1a), can be changed to the following form [1]:

$$\frac{\partial P}{\partial t} + v\frac{\partial P}{\partial x} + \rho a^2 \frac{\partial v}{\partial x} = 0$$

(2)

Where a is the acoustical wave velocity.

The Momentum equation, that is Eq. (1b), may be rearranged by multiplying A, the cross-sectional area of pipe to give

$$\frac{\partial(\rho v A)}{\partial t} + \frac{\partial(\rho v A v)}{\partial x} + \frac{\partial(P A)}{\partial x}$$
$$+ 2\rho A v^2 \left[\frac{\varphi_f}{D}\right] + \rho A g \sin\theta = 0$$

The first two terms of the above may be expanded as follows:

$$\frac{\partial(\rho v A)}{\partial t} + \frac{\partial(\rho v A v)}{\partial x} = v\frac{\partial(\rho A)}{\partial t} + \rho A\frac{\partial v}{\partial t} + v\frac{\partial(\rho v A)}{\partial x}$$
$$+ \rho v A\frac{\partial(v)}{\partial x}$$

Or

$$\frac{\partial(\rho v A)}{\partial t} + \frac{\partial(\rho v A v)}{\partial x} = v\left[\frac{\partial(\rho A)}{\partial t} + \frac{\partial(\rho v A)}{\partial x}\right]$$
$$+ \rho A\frac{\partial v}{\partial t} + \rho v A\frac{\partial(v)}{\partial x}$$

From the Continuity equation Eq. (1a), the term in bracket vanishes; therefore, the Momentum equation becomes

$$\rho\frac{\partial(v A)}{\partial t} + \rho v\frac{\partial(v A)}{\partial x} + \frac{\partial(P A)}{\partial x} + 2\rho A v^2 \left[\frac{\varphi_f}{D}\right]$$
$$+ \rho A g \sin\theta = 0$$

(3)

In most engineering applications, the convective acceleration terms, $v[\partial v/\partial x]; v[\partial P/\partial x]$ and the slope term are very small compared to the other terms in the above equations and may be neglected [1]. After dropping these terms from and, the following are obtained:

$$\frac{\partial P}{\partial t} + \rho a^2 \frac{\partial v}{\partial x} = 0 \tag{4}$$

$$\frac{\partial(\rho v)}{\partial t} + \frac{\partial P}{\partial x} + 2\rho v^2 \left[\frac{\varphi_f}{D}\right] = 0 \tag{5}$$

The discharge Q may be written as

$$Q = vA \tag{6}$$

Substituting Eq. (6) into and gives

$$\frac{\partial P}{\partial t} + \frac{\rho a^2}{A} \frac{\partial Q}{\partial x} = 0 \tag{7}$$

$$\frac{\partial Q}{\partial t} + \frac{A}{\rho} \frac{\partial P}{\partial x} + \frac{2\varphi_f Q|Q|}{DA} = 0 \tag{8}$$

Rearranging and, transient flow through the horizontal pipe can be represented by the following set of equations:

$$\frac{\partial Q}{\partial x} = -\frac{A}{\rho a^2} \frac{\partial P}{\partial t} \tag{9}$$

$$\frac{\partial P}{\partial x} = -\frac{\rho}{A} \frac{\partial Q}{\partial t} - \frac{2\varphi_f \rho Q|Q|}{DA^2} \tag{10}$$

Taking into account that

Mass flow rate $= \rho Q = \rho_n Q_n$

Where the subscript n refers to quantities at standard conditions of pressure $P_n \cong 0.1$ MPa and temperature $T_n = 288$ K.

The governing equations become

$$\frac{\partial Q_n}{\partial x} = -\frac{A}{\rho_n a^2} \frac{\partial P}{\partial t} \tag{11}$$

$$\frac{\partial P}{\partial x} = -\frac{\rho_n}{A} \frac{\partial Q_n}{\partial t} - \frac{2\varphi_f \rho_n Q_n|Q|}{DA^2} \tag{12}$$

Based on Eq. (11), we obtain

$$Q_n(x, t) - Q_n(x + \Delta x, t)$$

$$= Q_C = -\frac{\partial Q_n}{\partial x} \Delta x = \left[\frac{A\Delta x}{\rho_n a^2}\right]\left[\frac{\partial P}{\partial t}\right] \tag{13}$$

Where Q_c is the change in flow rate in a pipe dependent on compressibility of the fluid.

Therefore, for the consideration of capacitance effect in a pipeline network, the relationship between pressure and flow rate with capacitance can be made analogous to voltage and current relationship across an electric capacitor in the following form:

Nodal approach
$$J = G \frac{dV}{dt} \tag{14}$$

Mesh approach
$$V = \frac{1}{G} \int J \, dt \tag{15}$$

Where G is the capacitance, reflecting the capacitance effect in a pipeline. Comparing Eq. (14) with Eq. (13), the capacitance has the following form:

$$G = \frac{A \Delta x}{\rho_n a^2} = \frac{V_p}{\rho_n a^2} \tag{16}$$

Where V_p is the volume within the pipe.

For gas distributing system under isothermal conditions, we can write the equation of state in the form [8]

$$a^2 = \frac{P}{\rho} = \frac{z R_g T}{MW} \tag{17}$$

Substituting Eq. (17) into Eq. (16), we get

$$G = \frac{V_p MW}{\rho_n z R_g T} \tag{18}$$

Similarly, from Eq. (12), we obtain

$$P(x,t) - P(x + \Delta x, t) = -\Delta P = -\frac{\partial P}{\partial x} \Delta x = \frac{\rho_n \Delta x}{A} \frac{\partial Q_n}{\partial t} + \frac{2 \varphi_f \rho_n Q_n |Q| \Delta x}{D A^2}$$

(19)

We assume

$$-\Delta P_L = \frac{\rho_n \Delta x}{A} \frac{\partial Q_n}{\partial t} \tag{20}$$

And

$$-\Delta P_R = \frac{2\varphi_f \rho_n Q_n |Q| \Delta x}{D A^2} \qquad (21)$$

Where $(-\Delta P_L)$ is the pressure drop required to accelerate a given mass of fluid [6], which is due to inductance effect and is proportional to the rate of change of flow; and $(-\Delta P_R)$ is the pressure drop due to frictional resistance to flow. It is noted that Eq. (21) is the same as the Darcy–Weisbach formula. Therefore, the total pressure drop of a pipe is the sum of the pressure drop due to resistance effect and the pressure drop due to inductance effect across the pipe.

For the inductance effect in a pipeline network, based on Eq. (20) and the analogy between electrical and hydraulic flow, the analogous relationships for inductance relating to pressure drop and flow rate have the following forms, which are analogous to voltage and current relationship across an electric inductor:

Mesh approach $\qquad V = L \dfrac{dJ}{dt} \qquad (22)$

Nodal approach $\qquad J = \dfrac{1}{L} \displaystyle\int V \, dt \qquad (23)$

Where L is the inductance, reflecting the inductance effect in a pipe. Comparing Eq. (22) with Eq. (20), the inductance has the following form:

$$L = \frac{\rho_n \Delta x}{A} \qquad (24)$$

The mathematical representation of the resistance effect in a pipeline network has been employed in the steady state analysis. The adaptability of this model has already been demonstrated by many workers [2], [3], [10] and [11]. It can be expressed either in the forms of impedance Z or admittance Y, which correspond to the mesh or nodal approach. The relationship between the resistance with the pressure drop and flow rate is analogous to the Ohm's law and can be represented by the following equations:

Mesh approach $\qquad V = ZJ \qquad (25)$

Nodal approach $\qquad J = YV \qquad (26)$

The mathematical model for resistance in a pipeline network depends on the various equations describing the friction factor relationship between pressure drop and flow rate. Weymouth equation has been used in the previous work [10] on Osiadacz's sample networks [8]. Other equations used in the present study are listed as follows [8]:

Lacey's equation — (for the pressure range of 0–75 mbar gauge):

$$\varphi_f = 0.0044 \left[1 + \frac{12}{0.276D} \right]$$

(27a)

The Polyflo equation — (for the pressure range of 0.75–7.0 bar gauge):

$$\sqrt{\frac{1}{\varphi_f}} = 5.338 Re^{0.076} \eta$$

(27b)

The Panhandle 'A' equation — (for the pressure range above 7.0 bar gauge):

$$\sqrt{\frac{1}{\varphi_f}} = 6.872 Re^{0.073} \eta$$

(27c)

Where η is the efficiency factor accounting for the additional frictional or drag losses other than losses due to viscous forces.

DERIVATION OF EQUATIONS FOR THE TRANSFORMATION APPROACH

In order to derive the mathematical model for transient flow analysis, some of the fundamental assumptions used in the conventional methods are also applied. These assumptions are (i) one-dimensional and isothermal flow, and (ii) applying the steady state friction factor equation to transient flow [1] and [8].

As mentioned above, the change of flow rate across a pipe with time is due to compressibility of the fluid, i.e. the capacitance effect.

The pressure drop of a pipe is the sum of the pressure drop due to resistance effect and the pressure drop due to inductance effect across the pipe. A typical branch of fluid network with transient flow can be represented as an electrical circuit and be visualised in Fig. 1. The resistance and inductance are proposed to be connected in series, and the capacitance and resistance are proposed to be connected in parallel.

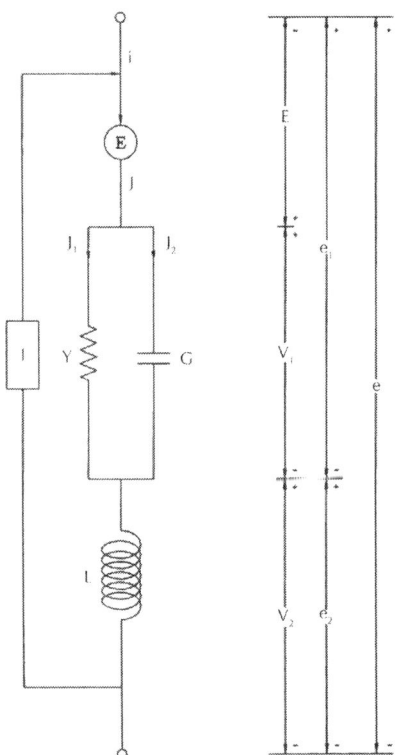

Figure 1: Composite branch of a network.

Based on Fig. 1, the following relationships can be drawn:

$$V = V_1 + V_2 \tag{28}$$

$$V_1 = E + e_1 \tag{29}$$

$$V_2 = e_2 \tag{30}$$

$$J = I + i \tag{31}$$

$$J = J_1 + J_2 \tag{32}$$

For the nodal approach [14]

$$J_1^b = Y^{bb}V_{1b} \tag{33}$$

$$J_2^b = G^{bb}\frac{dV_{1b}}{dt} \tag{34}$$

Based on Eq. (22),

$$V_{2b} = L_{bb}\frac{dJ^b}{dt} \tag{35}$$

Applying the transformation theory [14],

$$J^s = A_{\cdot b}^s J^b = A_{\cdot b}^s(J_1^b + J_2^b) = A_{\cdot b}^s\left[Y^{bb}V_{1b} + G^{bb}\frac{dV_{1b}}{dt}\right] \tag{36}$$

$$V_{2s} = C_s^{\cdot b}L_{bb}C_{\cdot s}^b\frac{dJ^s}{dt} \tag{37}$$

Substituting

$$V_{1b} = A_b^{\cdot s}V_{1s} \tag{38}$$

Into Eq. (36), the following equation is obtained:

$$\begin{aligned}
J^s &= A_{\cdot b}^s\left[Y^{bb}A_b^{\cdot s}V_{1s} + G^{bb}A_b^{\cdot s}\frac{dV_{1s}}{dt}\right] \\
&= A_{\cdot b}^s Y^{bb}A_b^{\cdot s}V_{1s} + A_{\cdot b}^s G^{bb}A_b^{\cdot s}\frac{dV_{1s}}{dt}
\end{aligned} \tag{39}$$

When   and   are extended to open path and closed path frameworks, they become

$$\begin{bmatrix} J^o \\ \overline{J^c} \end{bmatrix} = \begin{bmatrix} A_{\cdot b}^o \\ A_{\cdot b}^c \end{bmatrix} Y^{bb}[A_b^{\cdot o}|A_b^{\cdot c}]\begin{bmatrix} V_{1o} \\ V_{1c} \end{bmatrix}$$
$$+ \begin{bmatrix} A_{\cdot b}^o \\ A_{\cdot b}^c \end{bmatrix} G^{bb}[A_b^{\cdot o}|A_b^{\cdot c}]\begin{bmatrix} dV_{1o}/dt \\ dV_{1c}/dt \end{bmatrix} \tag{40}$$

$$\begin{bmatrix} V_{2o} \\ V_{2c} \end{bmatrix} = \begin{bmatrix} C_o^{\cdot b} \\ C_c^{\cdot b} \end{bmatrix} L_{bb}[C_{\cdot o}^b|C_{\cdot c}^b]\begin{bmatrix} dJ^o/dt \\ dJ^c/dt \end{bmatrix} \tag{41}$$

As

$$V_{1o} = E_o + e_{1o} = C_o^{\cdot b} E_b + e_{1o} \tag{42}$$

$$V_{1c} = E_c = C_c^{\cdot b} E_b \tag{43}$$

For an invariant pressure source E_b, we obtain the following relationships:

$$\frac{dV_{1o}}{dt} = \frac{de_{1o}}{dt} \tag{44}$$

$$\frac{dV_{1c}}{dt} = 0 \tag{45}$$

Expanding and, and incorporating and, we have

$$J^o = A_{\cdot b}^o Y^{bb} A_b^{\cdot o} (C_o^{\cdot b} E_b + e_{1o}) + A_{\cdot b}^o Y^{bb} A_b^{\cdot c} C_c^{\cdot b} E_b$$

$$+ A_{\cdot b}^o G^{bb} A_b^{\cdot o} \frac{de_{1o}}{dt} \tag{46}$$

$$J^c = A_{\cdot b}^c Y^{bb} A_b^{\cdot o} (C_o^{\cdot b} E_b + e_{1o}) + A_{\cdot b}^c Y^{bb} A_b^{\cdot c} C_c^{\cdot b} E_b$$

$$+ A_{\cdot b}^c G^{bb} A_b^{\cdot o} \frac{de_{1o}}{dt} \tag{47}$$

$$e_{2o} = C_o^{\cdot b} L_{bb} C_{\cdot o}^b \frac{dJ^o}{dt} + C_o^{\cdot b} L_{bb} C_{\cdot c}^b \frac{dJ^c}{dt} \tag{48}$$

Rearranging Eq. (46), we have

$$\frac{de_{1o}}{dt} = [A_{\cdot b}^o G^{bb} A_b^{\cdot o}]^{-1} [J^o - A_{\cdot b}^o Y^{bb} A_b^{\cdot o} (C_o^{\cdot b} E_b + e_{1o})$$

$$+ A_{\cdot b}^o Y^{bb} A_b^{\cdot c} C_c^{\cdot b} E_b] \tag{49}$$

, and are the governing equations for transient pipe flow system with a constant pressure source; and they are a set of first-order ordinary differential equations. Hence, the transient pipe flow problem which is ordinarily governed by the set of two partial differential equations can now be solved by a set of first-order ordinary differential equations which is much easier to handle. Nevertheless, these equations cannot be solved analytically owing to the complicated relationships involved in the network problem.

For a network with known topology, tensors $A_{.b}{}^{o}, A_{b}{}^{.o}, A_{.b}{}^{c}, A_{b}{}^{.c}, C_{.o}{}^{b}, C_{.o}{}^{b}, C_{.c}{}^{b}$, and $C_{c}{}^{.b}$ and inductance are independent of time and can be determined easily; while the boundary condition of admittance and capacitance can be calculated based on the results of steady state analysis. Therefore, the dynamic response of e_{1o} can be determined through the solution of a series of ordinary differential equations as Eq. (49). When e_{1o} is found, J^{c} can be calculated from Eq. (47). Then, e_{2o} is calculated from Eq. (48). Using e_{o} and J^{c}, the dynamic change of branch flow, branch pressure drop and nodal pressure of network at any given time can be found through the application of the transformation techniques.

COMPUTATIONAL SCHEME

Once the topology of the pipeline network is ascertained, the steady state analysis of pipeline network can be carried out to find the branch flow rate and nodal pressure, which are then used as the initial value to solve the ordinary differential equation. The steady state analysis of a pipe network can be based either on the mesh method or its dualistic nodal method. Once the steady state analysis is completed, the transient calculation is carried out by using the general solution method such as the fourth-order Runge–Kutta method for the first-order ordinary differential equations. The detailed computation scheme is outlined in Fig. 2. The computer program is written in Microsoft FORTRAN and runs on a Pentium/75 PC.

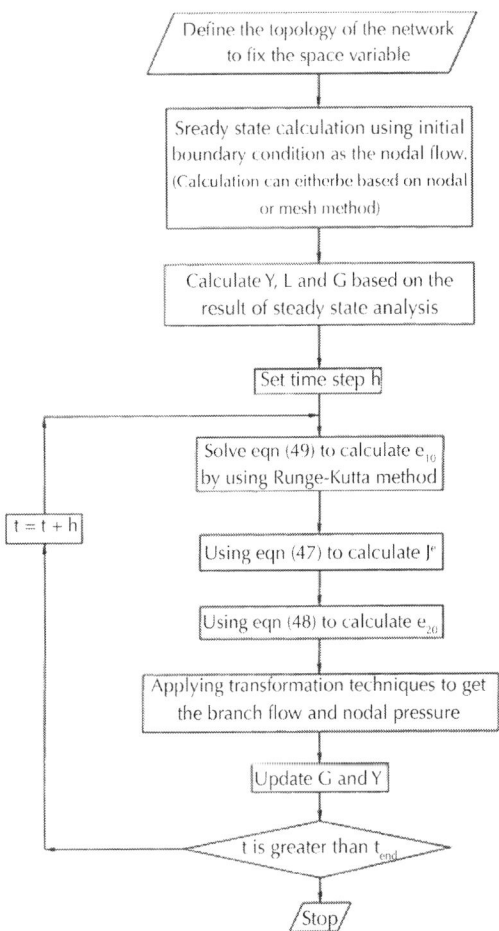

Figure 2: Computational flow chart.

SAMPLE NETWORKS

The mathematical model derived in the present study was tested on three simple gas networks, two of which have been analysed by Osiadacz [8] and one by London Research Station (LRS) [4]. The first sample network is a straight pipeline ($l = 10^5$ m) with a uniform diameter of 0.6 m. The upstream pressure is maintained at a constant level of 5 MPa. The flow rate is depicted in Fig. 3. The operating temperature is 278 K; the density of the fluid is 0.73 kg m⁻³; and the specific gravity of the fluid is 0.6. For the purpose of analysing, the pipeline is cut into five segments. Thus, the capacitance of each segment is five times the capacitance of the pipe. Osiadacz's result is presented in Fig. 4.

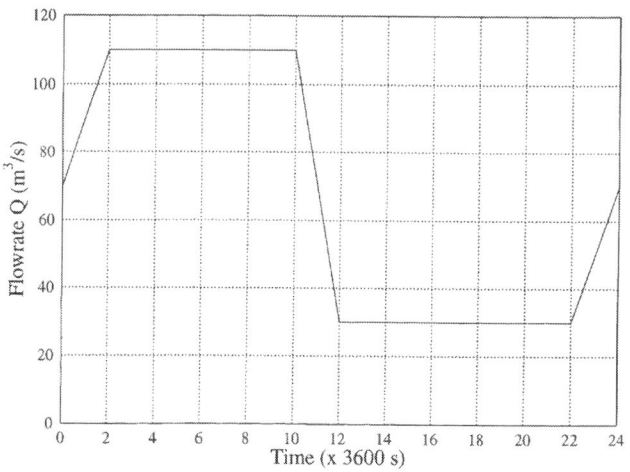

Figure 3: Change of flow with time (boundary condition).

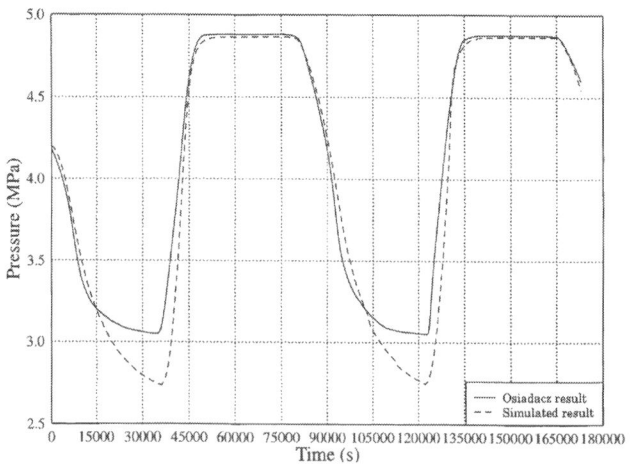

Figure 4: Change of pressure at the outlet for Sample Network 1.

The second example is a simple network with three nodes and three branches forming one mesh as shown in Fig. 5. The physical data are shown in Table 1.

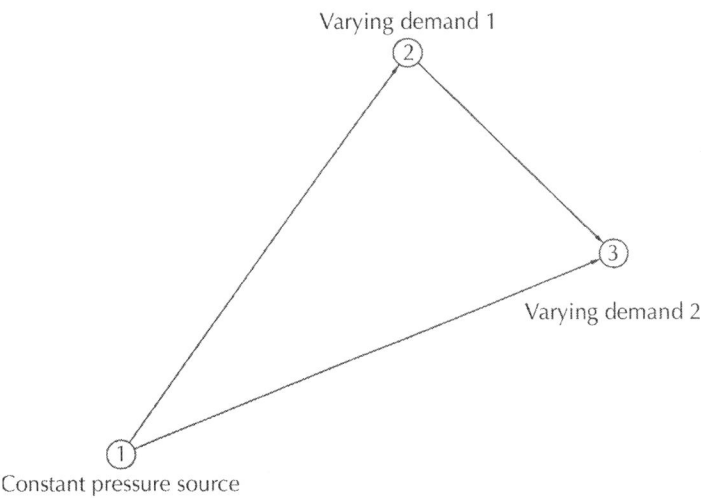

Figure 5: Sample Network 2.

Table 1: Pipe data of the second sample network[a]

Pipe	From	To	Diameter (m)	Length (m)
1	1	3	0.6	80000
2	1	2	0.6	90000
3	2	3	0.6	100000

a $=0.7165 \mathrm{kgm}^{-3}$; $S=0.6$; $T=278\mathrm{K}$; $t_{max}=86,400\mathrm{s}$.

Node 1 is the pressure source with a constant pressure of 5 MPa. The loads at Nodes 2 and 3 are known functions of time, which vary according to the curves depicted in Fig. 6. In this example, each pipe is cut into four segments. Hence, the capacitance of each segment is equal to the capacitance of the pipe multiplied by 4 [15].

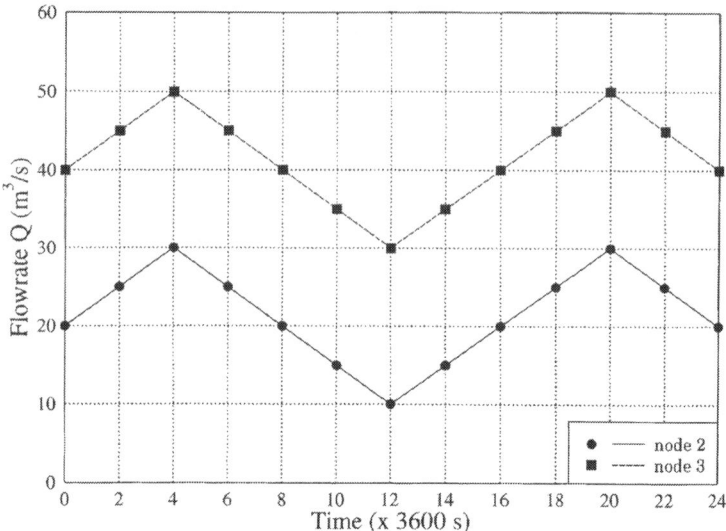

Figure 6: Changes of load at Nodes 2 and 3 for Sample Network 2.

The last sample network is a straight pipe having a uniform diameter. This sample network was used by LRS in demonstrating the versatility of their program PAN. Initially, the pipe was in a steady state; the demand then increased by 50% in a step change and was held constant thereafter. The pressure at the inlet, P_1, was held constant, while the pressure at the outlet, P_2, fell to a new steady state value. In this case, each pipe is cut into four segments and the capacitance of each segment is four times the capacitance of the pipe. The analysis was carried out for five different pressure ranges which corresponded to typical pressure range for low, medium and high. The operating conditions are listed in Table 2. The LRS results are depicted in Fig. 9, Fig. 10, Fig. 11, Fig. 12 and Fig. 13.

Table 2: Pipe data of the third sample network

	Length (mile)	Diameter (in.)	Initial flow rate (MSCFH)	Inlet pressure
LRS 1	80	18	1500	350 psig
LRS 2	20	12	800	180 psig
LRS 3	6	14	300	25 psig
LRS 4	3	14	80	2 psig
LRS 5	0.6	4	4	20 in. water gauge

The compressibility factor in this study is determined by using the following correlation:

$$z = \frac{1}{1 + \alpha P_{ave}}$$

(50)

Where α is obtained from linear interpolation of the constant taken from Gas engineers' Handbook [9] and P_{ave} is the average pressure of each pipe section.

RESULTS AND DISCUSSION

The dynamic pressure response at the outlet of the first sample network as simulated by the derived model is presented in Fig. 4. The computation time spent for solving the examples ranges from approximately 10 s to 5 min, depending on the different time steps used.

Simulation results of the second sample network using the present model is compared with the results from the literature. These are shown in Fig. 7 and Fig. 8.

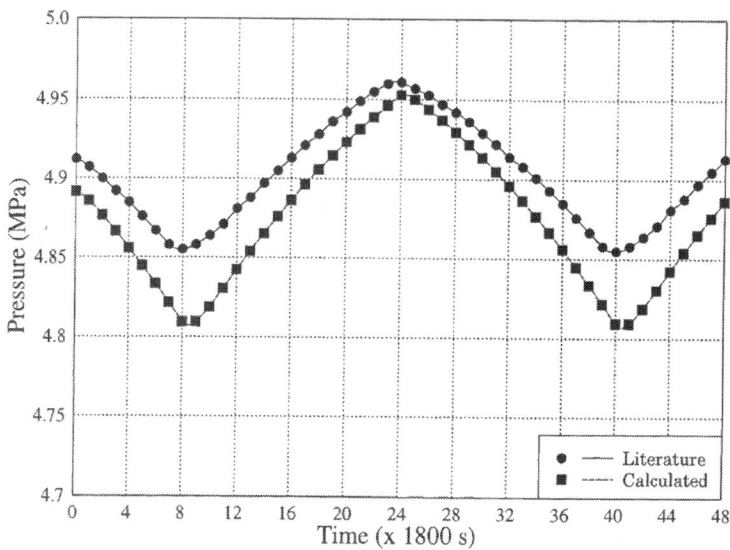

Figure 7: The variation of pressure at Node 2.

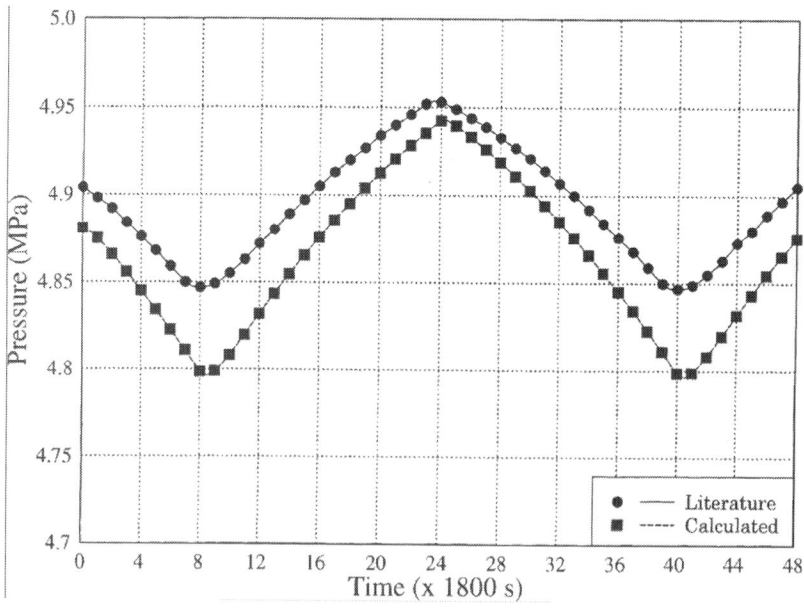

Figure 8: The variation of pressure at Node 3.

It can be seen that the results obtained by using the present method are comparable with those in the literature. The deviation is less than 9% for the first example and 1% for the second example. When the last sample network was analysed, it was found that the average simulated pressure response was higher than the LRS result by around an average of 10% (the maximum error is around 27% for the case of lowest pressure). This result shows that the response was slower, due to the high value of capacitance. Upon further investigation, it was found that, when the temperature was decreased, the capacitance became larger and the response slowed down. As most pipes were buried underground and the pipe wall temperature subjected to daily temperature change, the assumption of isothermal condition made in the derivation of Eq. (18) might not satisfy the LRS conditions. In order to avoid a more complicated model of considering the conservation of energy, an efficiency factor is adopted in the present study to consider the deviation from the assumption of isothermal flow condition. After numerous trials, a factor of 0.65 was selected based on the first case of LRS result (P_1 = 350 psig). It was found that the same factor of 0.65 would fit the other four cases of LRS with a maximum error of 7%. The results are depicted in Fig. 9, Fig. 10, Fig. 11, Fig. 12 and Fig. 13.

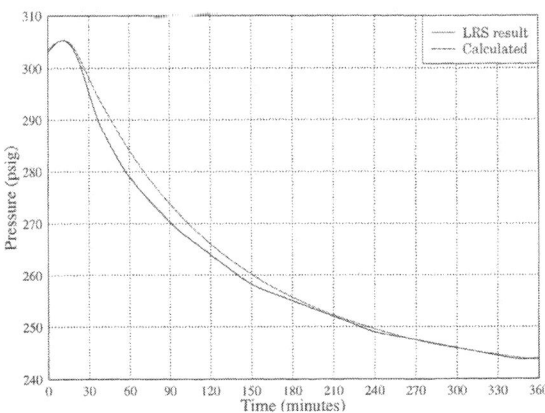

Figure 9: Comparison of simulated pressure with the LRS result (P_1 = 350 psig).

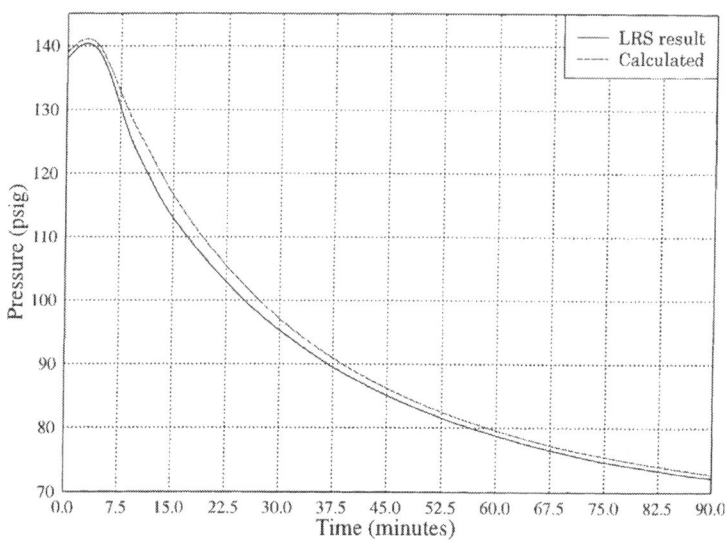

Figure 10: Comparison of simulated pressure with the LRS result (P_1 = 180 psig).

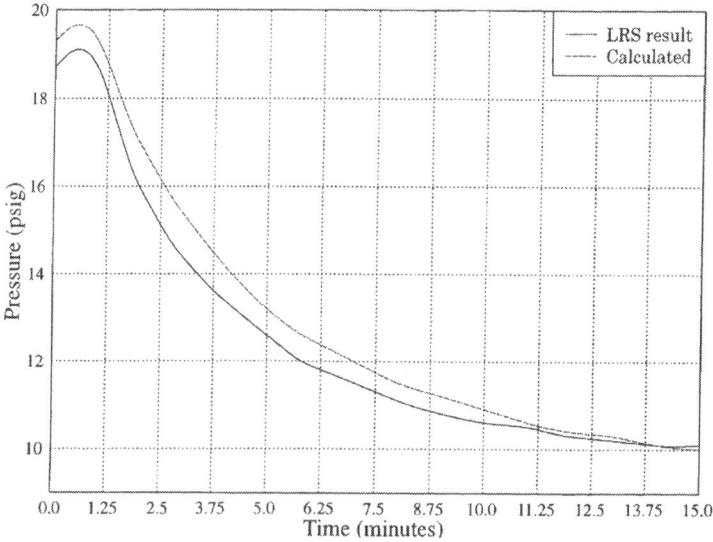

Figure 11: Comparison of simulated pressure with the LRS result (P_1 = 25 psig).

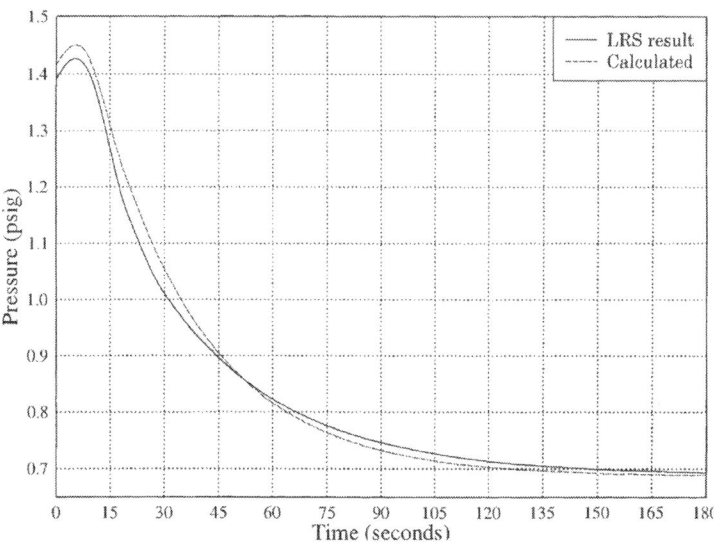

Figure 12: Comparison of simulated pressure with the LRS result (P_1 = 2 psig).

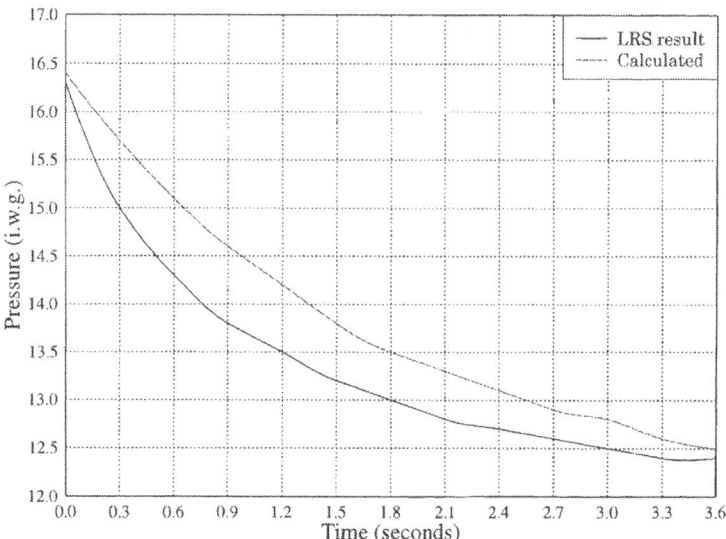

Figure 13: Comparison of simulated pressure with the LRS result (P_1 = 20 i.w.g.).

Convergence is normally a common problem in the analysis of pipeline network. However, in this present study, iteration process is only required in the part of steady state analysis; and the solution of transient analysis can be obtained without any convergence difficulty. Moreover, the steady state analysis was carried out using the robust transformation method [3] and [11], and hence, convergence problem was manageable.

The present study also investigated the effect of space and time discretization on the accuracy of solution. This is done by comparing the simulation results of the second sample network when all pipes were cut into four and eight equal parts. It is found that discretization of pipe section does not affect the precision of simulation result for the pipe network examined. It is also found that the variation of time step has negligible effect on the simulation result. On the other hand, both space and time discretization have noticeable effect on the computation effort needed, i.e. the finer the space or time steps, the more is the computation time required. For a node with a sudden change of high demand, a smaller time space is recommended.

The inductance effect on the solution of gas transient flow is also investigated. Simulation results of pipe flow with the inductance effect considered were compared to that when inductance effect was neglected. It is found that the effect of inductance which corresponds to kinetic energy of gas is negligible on the simulation results for the gas pipe networks examined compared to the resistance and capacitance effects. Hence, when the boundary conditions of a gas pipeline network do not change rapidly or the capacity of the pipe is relatively large, the effect of inductance can be neglected. Nevertheless, further study on the inductance effect for hydraulic transient analysis is recommended.

CONCLUSIONS

A new mathematical model based on electrical analogy and transformation theory is developed for transient analysis of isothermal gas flow in pipe networks. The transient behaviour of a pipe, conventionally taking the form of a set of second-order partial differential equations, can be expressed by a set of first-order ordinary differential equations using the new model. The solutions computed using the new method are compatible with those using the conventional methods. The new method is simple, straight forward and without convergence problem. It is easy to implement on a small personal computer. Comparing with the conventional methods, this new method shows a promising feature in saving computational efforts and may be employed as a powerful means for pipe network design and control.

REFERENCES

1. M.H. Chaudhry, Applied Hydraulic Transients, Van Nostrand Reinhold, New York, 1987.

2. B. Gay, P. Middleton, The solution of pipe network problems, Chem. Eng. Sci. 26 (1971) 109–123.

3. B. Gay, P.E. Preece, Matrix methods for the solution of fluid network problems Part I. Mesh methods, Trans. Inst. Chem. Engrs. 53 (1975) 12–15.

4. London Research Station, Proceedings of PAN Presentation, British Gas Corporation, 27 February 1974.

5. S.D. Millston, Electric analogies for hydraulic analysis Part 1. System components, Machine Design December (1952) 185–189.

6. S.D. Millston, Electric analogies for hydraulic analysis Part 2. Tubing characteristics, Machine Design January (1953) 166–170.

7. G. Murphy, D.J. Shippy, H.L. Luo, Engineering Analogies, Iowa State University Press, IA, 1963.

8. A.J. Osiadacz, Simulation and Analysis of Gas Networks, E. & F.N. Spon, London, 1987.

9. C.G. Segeler, M.D. Ringler, E.M. Kafka, Gas Engineers' Handbook, AGA, NY, 1969.

10. W.Q. Tao, H.C. Ti, Transient analysis of gas pipeline network, Chem. Eng. J. 69 (1997) 47–52.

11. H.C. Ti, K.G. Koh, H.M. Tar, Steady analysis of gas flow in a pipeline network, in: Proceedings of the 4th Asean Council on Petroleum (ASCOPE) Conference & Exhibition, November 1989, Singapore, and pp. 484–489.

12. J.F. Wilkinson, D.V. Holliday, E.H. Bakey, K.W. Hannah, Transient Flow in Natural Gas Transmission Systems, AGA, NY, 1965.

13. E.B. Wylie, V.L. Streeter, Fluid Transients, McGraw-Hill, New York, 1978.

14. G. Kron, Tensor Analysis of Networks, MacDonald, London, 1965.

15. R.G. Meadows, Electric Network Analysis, the Athlone Press, London, 1972.

Chapter 7

Turbulent Flow of Hydrates in a Pipeline of Complex Configuration

B.V. Balakin[a], A.C. Hoffmann[a], P. Kosinski[a], and
S. Høiland[b]

[a]Department of Physics and Technology, University of Bergen, Allegaten 55, 5007 Bergen, Norway
[b]SINTEF Petroleum Research, Thormøhlensgate 55, 5008 Bergen, Norway

ABSTRACT

The present paper describes an experimental study of Freon R11 hydrate transport in a turbulent flow through a model-scale pipeline of complex configuration. The frictional pressure losses as a function of the hydrate phase concentration were determined, and isokinetic sampling was performed for the determination of mean particle size. The experimental rig was modelled using an Eulerian–Eulerian CFD model which was validated with the experimental results. The detailed flow patterns in the pipeline and some other process parameters were determined by the CFD simulations, elucidating further the flow of the slurry in the pipeline. In addition the influence of gravity in densifying the particles under conditions of low Reynolds numbers was studied.

INTRODUCTION

Flow assurance under hydrate-stable conditions is an important scientific problem nowadays (Sloan and Koh, 2008) since the appearance of gas hydrates (Hashemi et al., 2009) in pipelines during oil processing may cause significant costs due to increased pressure loss or even plugging in the system. Laboratory experiments on pipelines containing hydrates require complicated high-pressure equipment, something which makes the visualization of flow patterns problematic. However, such experiments may yield information on relevant parameters such as hydrate slurry rheometry and mean particle size.

Two pioneering papers on high-pressure systems with natural gas hydrates do exist, however. The first paper by Darbouret et al. (2008) focuses on the study of hydrate slurry crystallization from a water-in-oil emulsion using chord-length measurements. During experiments using two different flow loops, a lab scale loop and a pilot-scale loop, the authors found, among other things, that the hydrate particle increase was the reason for a decrease of the slurry velocity. Darbouret et al. (2008) managed to distinguish between hydrate particle growth and their aggregation/dissociation on the basis of coupled pressure drop and chord length distribution (CLD) measurements. However, as stated by the authors, the measured CLDs do not provide the actual particle size distribution, this distribution may be recovered, however, by calculation. The second paper by Greaves et al. (2008) deals with hydrate formation in a high water cut system. The hydrate formation was studied for the closed agitated system with a specially designed digital camera and an FBRM-probe. Hydrate particle CLDs as well as *in situ* particle pictures were obtained during the work. It was shown in the paper that the hydrates in a water-in-oil emulsion can agglomerate rapidly, and therefore have the potential of forming an agglomerated plug in an industrial system.

The rest of the literature is, due to the above-mentioned complications with high-pressure systems, mainly focused on low-pressure hydrates.

Sinquin et al. (2004) studied the rheological properties of hydrate suspensions for laminar and turbulent flow regimes. They found for laminar flow that the apparent viscosity of the hydrate suspension increased proportionally to the hydrate volume fraction. For turbulent flow the hydrate–wall interaction was expressed in terms of the pipeline friction factor, which was also found to be dependent on the volume fraction of hydrate particles. They proposed an expression relating the hydrate agglomerate particle size to the fractal dimension, mean shear rate in the system, hydrate volume fraction and hydrate–hydrate adhesion force. However, the actual particle sizes were not determined experimentally in the work and conclusions on the validity of mentioned expression were drawn indirectly on basis of estimation of the suspension effective viscosity (which was assumed depend on the particle size) and comparing the pressure drop calculated on basis of this rheological expression with experimental results.

The problem of limited insight into the flow of hydrate suspensions obtainable from high-pressure experimental rigs may be resolved by performing experiments using so-called "low-pressure hydrates" which are formed by refrigerants such as freons and cyclopentane and also, in the past, by tetrahydrofuran. Some of these materials are widely used in the refrigeration industry (Wang et al., 2008).

Wang et al. (2008) studied the turbulent flow of HCFC hydrates in a low-pressure flow loop, involving the measurement of pressure gradients in the slurry and calculation of the pipeline friction factors. A series samples were obtained from the flowing suspension for determining the hydrate particle size distributions. A significant increase in the particle size was found as the hydrate concentration was increased. The actual rheology of the hydrate slurry and the detailed flow patterns of the suspension were not considered in this work.

Darbouret et al. (2005) studied the laminar flow of TBAB hydrate pulp at different temperatures. The effect of temperature on the system rheology was reported for two different types of TBAB hydrates, while the particle size was not determined in the work.

Computational fluid dynamics (CFD) studies may enhance the understanding of flow in hydrate-containing pipeline and therefore constitute a valuable tool in flow assurance research.

The range of CFD-focused research dealing gas hydrates is not very large at the present time and mainly involve simulations of static systems. For example, in the work by Sean et al. (2007) the process of hydrate dissociation in laminar flow was considered. The dissociation rate of a hydrate ball suspended in a flow cell was studied experimentally and the study supported by CFD simulations. The process of hydrate-particle movement with the fluid was not considered in the work. Moreover, the geometry studied was quite simple while pipeline networks in industrial systems are of more complex structures.

The dynamic process of methane hydrate deposition in a pipeline is considered in theoretical work byJassim et al. (2008) for different particle sizes and compared with experimental data. The CFD-model was one-way-coupled, i.e. the carrier phase flow was initially simulated with CFD. After that the in-house mechanistical model for particles deposition was applied using pre-computed velocity filed.

The present paper considers the process of low-pressure Freon R11 hydrates transport in the turbulent flow of complex configuration. A set of experiments was conducted for clarification of the slurry rheology, and flow sampling was carried out with subsequent further microscopical studies for finding particle sizes at different hydrate concentrations. An Eulerian–Eulerian CFD model of the experimental rig was generated for the simulation of flow patterns. The model was turbulent, three-dimensional and two-way coupled allowing prediction of those details of the process which could not be ascertained experimentally.

METHODOLOGY

Flow-Loop Experiments

The hydrate slurry used in the current work was generated by addition of Freon R11 (trichlorofluoromethane) to tap water. This refrigerant is soluble in water with the solubility of 1.1 g/dm³ at 20°C. Its hydrates are of structure II and they are formed at atmospheric pressure and temperatures below 8.5°C (Berge et al., 1998).

The experimental rig was mounted in the cooling cabinet in order to keep the process temperature in the range of hydrate stability, so the piping dimensions were limited by a cabinet volume. The flow loop designed for that purpose consisted of (see Fig. 1):

- transparent PVC-pipe (length 2.3 m, diameter 4.52 cm);
- two 90° pipe curvatures with major radii of 165 mm;
- three 90° bends with major radii of 60 mm;
- a DN40 centrifugal pump (Froster);
- a coriolis mass flowmeter KCM 40 K (from KEM Küppers);
- three static pressure sensors GS 4200 (from GENSPEC);
- a thermocouple SEM 203P (from Status Instruments);
- valve system for freon injection, ventilation, sampling and draining procedures.

Samples of the suspension from the loop were studied with a stereoscopic zoom microscope SMZ800 (from Nikon). Hydrate particle pictures were obtained using a Retiga EXi digital camera (resolution 1392×1040), connected to the microscope. The experiments were performed under the conditions:

- static pressure on the sensors in the interval 1.0–1.24 bar;
- water temperature 2.0±1.2 °C.

The amount of hydrate-former (freon) in the loop was increased incrementally within the interval 0–934 ml in steps of ~100ml. For each increment of hydrate-former volume in the system, the mean flow velocity was varied in the range of 0–3.9 m/s; pressure drops and density measurements were recorded.

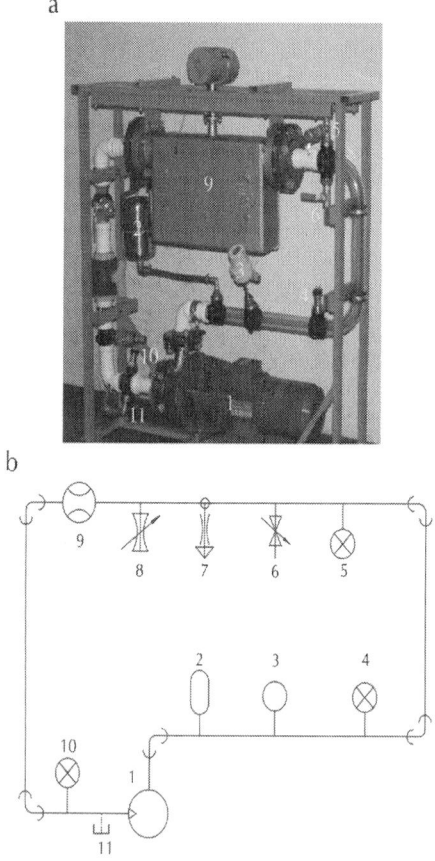

Figure 1: (a) photo of the experimental flow loop. (b) flow loop diagram (1—centrifugal pump; 2—expansion tank; 3—thermocouple; 4,5—pressure sensors; 6—sampling unit; 7—ventilation; 8—safety valve; 9—flow meter; 10—pressure sensor; 11—drainage).

Samples were taken isokinetically directly from the flowing suspension for further analysis. Fast and delicate further treatment of samples was conducted in order to avoid possible particle size changes associated with growth, aggregation, dissociation and breakage:

- The samples were immediately removed from the slurry by opening a valve, allowing the suspension to flow a beaker by gravity. The quick sampling process inducing a low shear rate compared to that than in the loop allows us to assume that the particle size change due to aggregation and breakage during sampling is negligibly low.

- The sampling beaker was placed in the same cooling cabinet as the loop in order to keep possible dissociation of particles to a minimum.

- The sample in the beaker was then made homogeneous by stirring with a magnetic stirrer (the duration of the stirring was no longer than 2–3 s). Again, the short duration of the stirring and the relatively low shearing rate induced would not lead to a significant particle size change; especially since it was found on a similar refrigerant-hydrate system that this type of hydrate is not very cohesive in water (Balakin et al., 2010).

- Photographs of hydrate particles in the homogeneous suspensions were taken afterwards in the same cooling cabinet.

During the entire process particle size change due to growth (not agglomeration but further crystallization on the particle surface) is assumed to be low since each sampling was done after the freon conversion into hydrate in the loop had reached completion.

It has to be noted that the particle—or agglomerate—size at any given time is not only the result of processes taking place in the pipeline. We have discussed this issue in our previous work (Balakin et al., 2010), where we modelled theoretically the particle size in a similar flow loop containing low-pressure hydrate slurry. During population balance modelling it was found that the processes taking place in the pump are also important for particle aggregation and breakage, and may either decrease or increase the system-average particle size dependent on the level of agitation in the pump. We would also like to stress that some dissociation of particles may have been caused by the heating effect of the pump, since the temperature in the rig was increased by 1.2°C at the maximum pump capacity. However, we attempted to minimize this problem by setting the impeller speed to its maximum value only during short periods during the experiments. More detailed descriptions of the experimental procedures may be found in Balakin et al. (2010).

CFD Model

The computational grid for numerical simulation of the experimental process was built using 66 363 polyhedral control volumes in STAR-design. The 3D CAD-model presented in Fig. 2 reproduces the part of experimental rig (Fig. 1) between the pump and the flow meter, which is considered as the test section. The model is made in real dimensions and includes some details of the flow loop geometry namely the three flow loop bends, the straight pipeline regions and the temperature sensor probe.

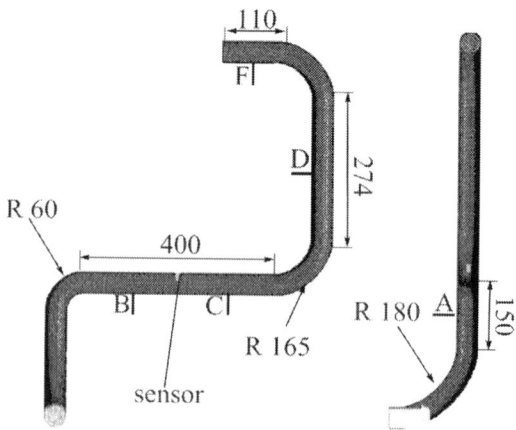

Figure 2: Three-dimensional CAD-model of the flow loop section between pump and flowmeter. Dimensions in mm.

The inlet boundary condition for the CFD simulations needed some care. The centrifugal pump may give rise to a non-uniform flow through the inlet to the test section (Cheah et al., 2007), since the total length of straight pipeline after the pump is not enough to establish fully developed turbulent flow. Using this idealized profile at the inlet would therefore compromise the accuracy of the simulation of the experimental process. In order to simulate the possible effect of the pump a curved pipe section was added to the computational domain upstream of the actual test section geometry.

The mesh was equipped with a four-layer prismatic subsurface in the neighborhood of the wall used for accurate calculation of the log-law velocity profile for the cases where the standard wall functions (Methodology, 2006) were applied in the turbulence model.

The carrier phase density and viscosity used in the simulation were 1000 kg/m^3 and 0.001 Pa s, respectively. The dispersed phase (hydrate) density used was 1138 kg/m^3. The apparent viscosity of the dispersed phase was set to depend on the volume fraction in a way described later in the paper. The hydrate-former properties were not taken into account in the simulation, since the hydrate former was assumed to be completely consumed before each series of pressure-drop measurements was conducted. The particle size used in the simulation was set to the value determined experimentally for the given concentration (see Table 2 further on in the paper).

The mean flow velocity and mean volumetric hydrate-phase concentration in the system were varied during the simulations. These parameters were set to those determined experimentally during the loop tests.

Multiphase flow was treated using the Eulerian–Eulerian technique (Pfleger et al., 1999), the flow is considered to be turbulent, isothermal, with no interphase mass transfer.

The governing equations are the continuity and momentum balances for each phase. The continuity equation is

$$\frac{\partial(\phi_m \rho_m)}{\partial t} + \nabla(\phi_m \rho_m \vec{u}_m) = 0 \tag{1}$$

with the constraint

$$\sum_{m=1}^{s} \phi_m(x,y,z) = 1.0 \tag{2}$$

In these equations $m=1$ or s indicates the phase, ϕ and ρ are the volume fraction and density, respectively, and $\vec{u}$ is the velocity vector. The momentum equation is

$$\frac{\partial(\phi_m \rho_m \vec{u}_m)}{\partial t} + \nabla(\phi_m \rho_m \vec{u}_m \vec{u}_m)$$
$$= -\phi_m \nabla p + \phi_m \rho_m \vec{g} + \nabla \cdot \phi_m(\underline{\underline{\tau_m}} + \underline{\underline{\tau_m^t}}) + \vec{M}_m \tag{3}$$

where p is the pressure, $\vec{g}$ the gravitational acceleration and $\underline{\underline{\tau}}$ and $\underline{\underline{\tau}}^t$ are the molecular and turbulent stress tensors, respectively. The interphase momentum-transfer terms are symmetric between the phases, such that $\vec{M}_l = -\vec{M}_s$ representing the drag, lift and virtual mass and $\vec{F}_{A,l} = -\vec{F}_{A,S}$ representing the buoyancy.

The viscous stress tensor for the liquid phase is given by (see, for example, Ferziger and Peric, 2001)

$$\tau_l^{ij} = 2\mu_l D_l^{ij} - \frac{2}{3}\mu_l \delta^{ij} \frac{\partial u_l^i}{\partial x^i} \tag{4}$$

where $D_l^{ij} = \frac{1}{2}\left(\partial u_l^i / \partial x^j + \partial u_l^j / \partial x^i\right)$ is the rate of strain tensor and the superscripts denote the contravariant vector and tensor components.

The stress tensor induced in the solid phase, τ_S^{ij}, is modelled using an expression similar to Eq. (4):

$$\tau_s^{ij} = 2\mu_s(\phi_s)D_s^{ij} - \frac{2}{3}\mu_s\delta^{ij}\frac{\partial u_s^i}{\partial x^i}$$

(5)

where μ_s is a "solids viscosity", a notion that we will be discussing further below.

The suspension viscosity increases with increasing solid phase concentration. In our case, the viscosity of the carrier phase (water), μ_l, is taken to be constant, independent of the solid phase concentration in the system. In order to account for the variation in the rheological behaviour of the entire suspension with the solids concentration we modify the viscosity of the solid phase (hydrate), μ_s, such that the resulting viscosity of the suspension approximately agrees with that determined experimentally. This was done by fitting the experimentally determined suspension viscosity by a relation of the form:

$$\frac{\mu_{susp}}{\mu_l} = (1-\phi_s)^a$$

(6)

A good fit was obtained for $a = -2.55$. Assuming that the viscosity of the whole suspension is approximately given by, $\phi_s\mu_s + \phi_l\mu_l$ the solid viscosity needed to give the fitted suspension viscosity can be found from

$$\mu_s = \frac{\mu_l((1-\phi_s)^a - \phi_l)}{\phi_s}$$

(7)

It has to be also noted that the rheological expression, used in the current work, does not include the value of the packing limit. It is possible to fit the rheological dependence, obtained in the experiments, by an expression

$$\frac{\mu_{susp}}{\mu_l} = \left(1 - \frac{\phi_s}{\phi_{max}}\right)^{a\cdot\phi_{max}}$$

(8)

where $\phi_{max} = 0.55$ is the packing limit value, determined experimentally (Balakin et al., 2010) and $a = -2.0$. However, the simulation results, done with the current expression, are in practice similar to one, produced without the use of the packing limit in rheological expression, for homogeneous flow regime.

The dependence of the solids viscosity on its concentration is presented in Fig. 3 together with the viscosity of the entire suspension which results from using Eq. (7).

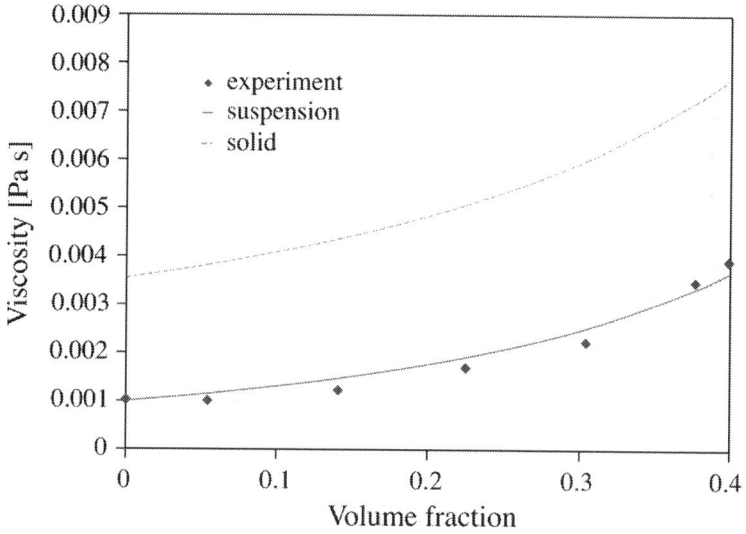

Figure 3: Dynamic viscosity as a function of hydrate volume fraction.

We justify the above procedure as follows: We note that the concept of a "solids viscosity" derives from the fundamental notion of two interpenetrating and interacting fluid phases, which are both continuous such that a shared pressure acts through each. This notion is the foundation of the multiphase momentum equations according to Gidaspow's model "A" (Gidaspow, 1994). In our case we do not have such a system, but a dispersed solids phase in a continuous fluid phase. Such a system is more naturally modelled using Gidaspow's model "B" equations. Nevertheless, Gidaspow showed that also with model A the fundamental equations for solid–fluid systems, e.g. fluidized beds, could be recovered, such that the dissipative flow force acting on the solid particles in a fluidized bed can be shown to balance the buoyant weight of the particles. Furthermore, experience shows that the two formulations yield almost identical simulation results in a wide variety of systems. Moreover, the current versions of STAR-CD left us the only the option of using the equations of model A.

Based on these arguments we decided the way to simulate the experimental system was to tweak the "solids viscosity" such that the resulting viscosity of the simulated suspension was in some sense equivalent to the experimental suspension viscosity, as we did through the above procedure. We thus do not attribute any direct physical meaning to the "solids viscosity", but consider it an artifact brought in by the use of Gidaspow's model A; one, however, that can be used to model the influence of the presence of the solids on the suspension viscosity.

Since the solids generally are suspended in our application, and the particle–particle interaction is weak, such as it is, for example, in a fluidized bed (Gidaspow, 1994) we have not included a term for the solids pressure in the solids momentum equation. Such a term, when used, is intended to account for particle–particle interaction giving rise to a "solids pressure". We assume that all the effects of interparticle interaction are accounted for by the terms for the viscous (Eqs. (5), (8) and (7)) and turbulent (Eq. (13)) stresses in the solid phase momentum equation.

The interphase momentum transfer term $\vec{M}$, the most important part of which is the drag force acting between the solids and liquid phases per unit volume of suspension, was modelled using the drag function:

$$\vec{F}_D = c_{ls}\vec{u}_r \qquad (9)$$

where $\vec{F}_D$ is the drag force acting between the phases per unit volume of suspension, and $\vec{u}_r$ is the relative velocity between the phases. For $\phi_s > 0.5$, the drag function, c_{ls}, is calculated according to the Ergun equation (Ergun, 1952 and Holbeach and Davidson, 2009):

$$c_{ls} = 150 \frac{\phi_s^2 \mu_l}{\phi_l d^2} + 1.75 \frac{\phi_s \rho_l}{d} |\vec{u}_r|.$$

(10)

For $\phi_s < 0.5 c_{ls}$ is calculated by the relation of di Felice (Crowe et al., 1998), which involves a correction factor $\phi_l^{-1.7}$ accounting for the effect on the drag on a particle of the presence of the other particles:

$$c_{ls} = \frac{3C_D}{4d} \phi_s \rho_l |\vec{u}_r| \phi_l^{-1.7}$$

(11)

The drag coefficient, C_D, is, for $Re_p \equiv \rho_l u_r d / \mu < 1000$, given by (Crowe et al., 1998)

$$C_D = \frac{24}{Re_p}(1 + 0.15 Re_p^{0.687})$$

(12)

When $Re_p > 1000$ the drag coefficient is set to 0.44 in accordance with Newton's drag law.

The turbulent stress $\underset{=}{\tau_m^t}$ is calculated with the eddy viscosity technique (Methodology, 2006)

$$\tau_m^{ijt} = \mu_m^t \left(2D_m^{ij} - \frac{2}{3}\delta^{ij}\frac{\partial u_m^i}{\partial x^i} \right) - \frac{2}{3}\rho_m k_m \delta^{ij}$$

(13)

The turbulent viscosity is in the carrier phase calculated as (Methodology, 2006)

$$\mu_l^t = C_\mu \rho_l \frac{k_l^2}{\varepsilon_l}$$

(14)

where $C_\mu = 0.09$. The turbulent viscosity for the solid phase is proportional to the one calculated for the liquid phase by Eq. (14):

$$\mu_s^t = \left(\frac{\rho_s}{\rho_l} \right) \mu_l^t$$

(15)

The transport equations for the turbulent kinetic energy, k_l, and the rate of turbulent energy dissipation,ε_l, per unit volume of suspension are (Methodology, 2006):

$$\frac{\partial(\phi_l\rho_l k_l)}{\partial t} + \nabla(\phi_l\rho_l\vec{u}_l k_l) = \nabla\left(\frac{\phi_l(\mu_l+\mu_l^t)\nabla k_l}{\sigma_k}\right) + \phi_l k_l(G-\rho_l\varepsilon_l)$$

(16)

$$\frac{\partial(\phi_l\rho_l\varepsilon_l)}{\partial t} + \nabla(\phi_l\rho_l\vec{u}_l\varepsilon_l) = \nabla\left(\frac{\phi_l(\mu_l+\mu_l^t)\nabla\varepsilon_l}{\sigma_\varepsilon}\right) + \phi_l\frac{\varepsilon_l}{k_l}(C_1 G-C_2\rho_l\varepsilon_l)$$

(17)

where

$$\underline{\underline{\tau_m^t}} G = \mu_l\left(\nabla\vec{u}_l +(\nabla\vec{u}_l)^\top\right): \nabla\vec{u}_l, C_1 = 1.44, C_2 = 1.92, \sigma_k = 1.0, \sigma_\varepsilon = \kappa^2 /\left[(C_1-C_2)\sqrt{C_\mu}\right]$$

with $\kappa = 0.4187$

The boundary conditions and initial conditions used in the model were as follows. The inlet boundary had prescribed k and ε, uniform velocities for both phases, zero interphase slip and a prescribed solid phase volume fraction. The outlet boundary condition was the standard one. At the solid walls no-slip boundary conditions were used for both phases using the standard wall functions (Methodology, 2006). The initial velocity field was set to zero and the pressure to atmospheric pressure for every simulation. The volume fraction of hydrate was set to be equal to the value measure experimentally using the coriolis flow meter.

Eqs. (1), (2), (3), (4), (5), (8), (7), (8), (9), (10), (11), (12), (13), (14), (15), (16) and (17) were discretized by the upwind scheme and solved iteratively with the semi-implicit method for pressure-linked equations (SIMPLE Patankar and Joseph, 2001). The simulation was considered to be converged when the maximum residual value was not higher than 1×10^{-5}.

RESULTS AND DISCUSSION

Experimental Results

After each injection of freon into the system it took only a short time for the hydrate former to be converted completely into hydrates.

This fact was confirmed by observing the variation of the system pressure during hydrate formation, and may be explained by the combination of two factors enhancing hydrate particle nucleation and growth:

- the subcooling of the system was about 6.5°C and
- the pump rotation frequency was set to a maximum value during hydrate growth,

the latter factor increasing the diffusion rate of hydrate-former. In the highly agitated system the first hydrate particles were visually observed around 20 s after the pump was turned on. When the process of freon consumption process was completed, the resulting volume fraction of particles was determined by a density measurement in the coriolis flowmeter. The hydrate phase volume fractions produced by the freon injections are presented in Table 1. It can be seen that the measured volume fraction of hydrate increases almost linearly with the injected volume of freon, as expected.

Table 1: Volume of freon injected and resulting volume fraction of hydrate

Volume of freon injected (ml)	Hydrate volume fraction
100	0.050
199	0.010
297	0.140
393	0.190
489	0.220
584	0.270
677	0.304
770	0.341
853	0.378
934	0.399

Hydrate volume fractions in the interval 0.05–0.1 did not affect the measured pressure drop much. At mean flow velocities in the interval 0.4–3.9 m/s (corresponding to Reynolds numbers of

18 080–176 280) it was observed visually that the concentration of particles was approximately uniform over most of the pipeline while minor variations in the appearance of the slurry were seen near the bends of the loop. However, when the mean velocity was decreased below 0.4 m/s (the flow was still turbulent) particles began to settle, producing visible regions of higher concentration in the bottom of the horizontal segments of the pipeline. The settled particles formed a bed in the horizontal regions of the loop even in the turbulent flow regime. CFD-simulations of hydrate bed behaviour in this region are the topic for our next study, the details are available from authors. Here we would like to note that a laminar flow regime was not achieved in the rig since the sensitivity of the pump frequency setting was not high enough to establish sufficiently small flow rates.

As the hydrate volume fraction increased to above 0.14, the pressure drops over the monitored sections started to increase above those measured for pure water flow. The slurry was blocked in the pipeline at hydrate concentration around 40%. The blockage was not caused by the formation of a plug-like obstacle by cohesive particles, but was the consequence of the overall frictional pressure loss increasing to above the characteristic curve of the pump. The absence of a plug-like obstacle, the formation of which might lead to plugging at much lower hydrate volume fractions, may be due to the low ad- and cohesion forces between the "cold" hydrate particles and the smooth plastic wall on the one hand and the other particles on the other hand Balakin et al. (2010). Also the powerful effect of the pump may prevent the formation of a cohesive obstacle.

The principle of "equivalent length" (Darby, 2001) was chosen to report the pressure drop data. This consists in calculating the length of a pipe of the same diameter as the loop piping but without any bends, fixtures or fittings generating the same pressure drop as the actual loop. The length of straight pipe generating a given pressure drop can be found from the Darcy–Weisbach equation (Viessman and Hammer, 2004 and Huang et al., 1994):

$$L_{equiv} = \frac{2d\Delta p}{\lambda \rho_w U_{mean}^2}$$

(18)

where Δp is the experimental pressure drop, U_{mean} is the mean water velocity and λ the dimensionless Blasius or Darcy friction factor, which can be calculated from the Blasius equation (de Nevers, 1991):

$$\lambda = \frac{0.3164}{Re^{0.25}}$$

(19)

The experimental pressure drops used to calculate L_{equiv} from Eq. (18) were those measured with pure water flowing in the system with Re in the interval from 3×10^3 to 2×10^5. The average value of L_{equiv} of 26.4 ± 7.8 m was used. Pressure gradients, calculated as $dp/dl=\Delta p/L_{equiv}$ are reported.

Average measured pressure gradients obtained using the equivalent length technique are shown as points plotted against the mean flow velocity (determined independently on the pump and flowmeter) for hydrate volume fractions of 0.00, 0.140, 0.220, 0.304 and 0.378 in Fig. 4. This figure also shows CFD results, which will be discussed later. The plots show that the pressure gradient in the pipeline increases with the mean flow velocity in agreement with Darcy equation (Viessman and Hammer, 2004). An increase in the pressure gradient with increasing solids concentration at constant flow rate is also evident, physically this is due to the increased apparent viscosity of the suspension with increasing solids concentration.

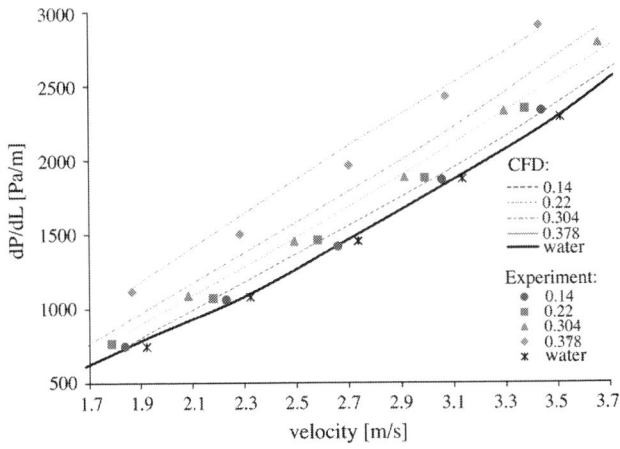

Figure 4: Pressure gradient as a function of mean flow velocity for hydrate volume fractions 0, 0.14, 0.22, 0.304 and 0.378. Experimental measurements are compared with CFD predictions for hydrate particle size 70μm.

Isokinetic sampling from the multiphase flow at a mean pipe velocity of 2 m/s made it possible to obtain micropictures of hydrate particles, such as the one presented in Fig. 5.

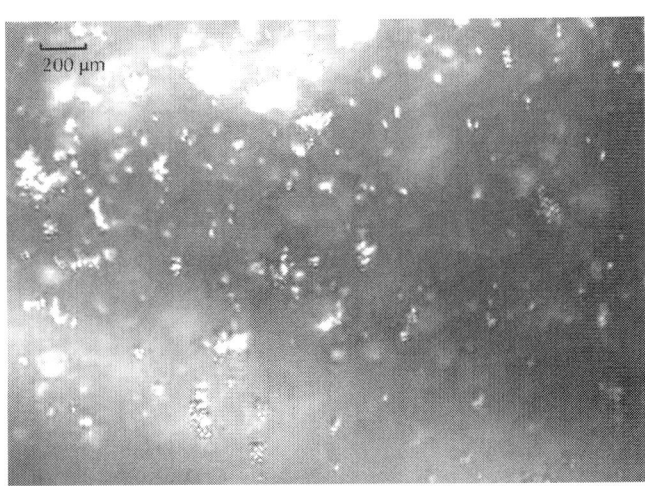

Figure 5: Microphotograph of Freon R11 hydrate particles, zoom3X Retiga EXi.

Particle size determination was done with the help of the commercial image analysis software ImagePro and the results for the mean particles size are presented for hydrate volume fractions 0.05, 0.1 and 0.14 inTable 2. The images of homogenized samples with larger volume fractions of hydrates were overcrowded with particles which made the separate particles difficult to detect on the background of a densely packed slurry. Thus the pictures were not sufficiently resolved for the ImagePro software to be able to identify separate particles on them. Dilution of the concentrated suspension samples with water would have caused an uncontrolled change in the particle size.

Table 2: Mean particle size as a function of hydrate volume fraction

Hydrate volume fraction	Mean particle size (µm)
0.05	27.0
0.10	48.0
0.14	70.0

From the table it can be seen that an increase of mean particle size was observed with the increase of hydrate volume fraction in the system, which is explained by both growth of the individual particles with every new freon injection and more intensive aggregation due to the higher concentration of particles.

CFD Modelling

To gain further understanding of the flow process and due to the difficulties in determining concentration profiles experimentally, related to the restrictions associated with fitting the equipment into the cooling cabinet, concentration profiles were predicted by CFD.

Simulations were conducted with the modelling approach described in Section 2.2. Model pressure gradients were determined for the entire set of hydrate concentrations and a constant particle size of 70µm for validation with the pressure gradients found experimentally. Detailed CFD profiles for flow loop velocity, hydrate concentration, particle Reynolds number and turbulent energy dissipation-rate were found for one test case at a hydrate volume fraction of 0.14 and a particle size of 70µm. Additional simulations were performed for a set of hydrate volume fractions and flow velocities to clarify the dependence of the particle Reynolds number on these two parameters.

In addition the process of hydrate phase gravitational densification on a low Reynolds numbers was simulated for hydrate volume fractions of 0.14 and 0.30.

The simulated pressure gradient is plotted against the mean flow velocity in Fig. 4 for hydrate volume fractions of 0.14, 0.22, 0.304 and 0.378 together with the experimental results. It can be seen from the figure that frictional pressure gradient in the model increases with hydrate volume fraction. The CFD-model predictions agree with experimental data well for 0 and 0.14 hydrate volume fractions in the system. At volume fractions greater than 0.14 the model slightly overpredicts experimental pressure gradients, although the discrepancy is less than 8%. Slight oscillations in the pressure drop predictions are explained by the effect of the secondary flowpatterns just downstream of the second bend, where the pressure tapping was located. Flow velocity magnitude profiles of the carrier phase established at a steady-state are presented in Figs. 6A, B for mean flow velocities 1.84 and 3.82 m/s (corresponding to pipe Reynolds numbers of 83 168 and 172 664). From the figures it can be seen that the zones of maximum velocity are associated with the inner part of the bends. The bends are too smooth to generate actual recirculation-zones in the outer part of the bends, but the flow is still retarded there. The temperature sensor probe indicated located in the lower horizontal part shifts the position of maximum velocity from a top of the line to the bottom.

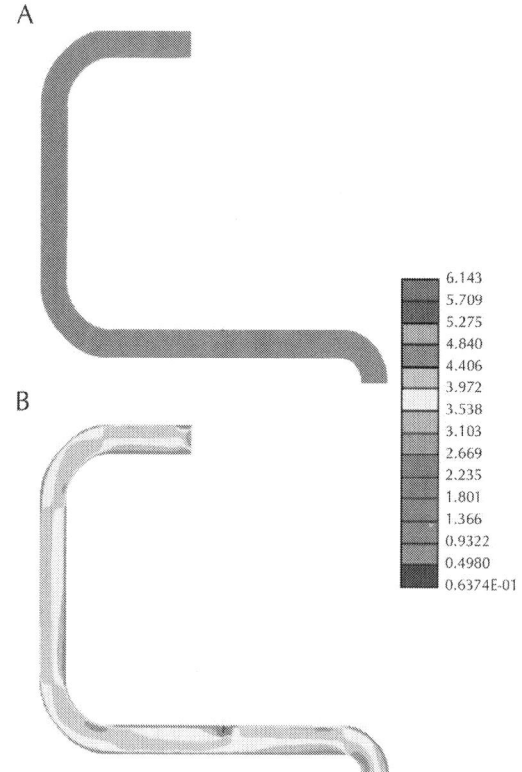

Figure 6: (A, B) Contour plots of the carrier phase velocity magnitude in the midline cross-section. Mean flow velocities 1.84 and 3.82 m/s, respectively, hydrate volume fraction: 0.14, hydrate particle size: 70μm.

An increase in the velocity (from A to B in the figure) leads to sharper gradients in the velocity magnitude; the positions of velocity maxima and minima are largely unchanged.

The volume fraction profiles (Figs. 7A, B) are mainly uniform over the loop midline cross-section. The variations observed visually during the experiments are reproduced by the CFD model, these are mainly slight minima in the particle concentration in the inner sections of the bends, just downstream of the bends. These are due to the particles being centrifuged outward in the bends. The variations are qualitatively similar for the two velocities, but stronger for the higher velocity, as expected.

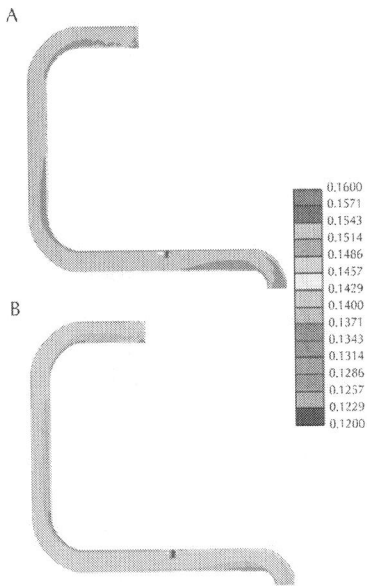

Figure 7: (A, B) Contour plots of hydrate volume fraction in the midline cross-section. Mean flow velocities 1.84 and 3.82 m/s, respectively, hydrate volume fraction: 0.14, hydrate particle size: 70μm .

Contour plots of the carrier-phase velocity magnitude are presented in Figs. 8A–F for the cross-sections indicated in Fig. 2. The cross-sections are selected so that they lie in the middle between local flow obstacles: inlet curvature, bends and temperature sensor.

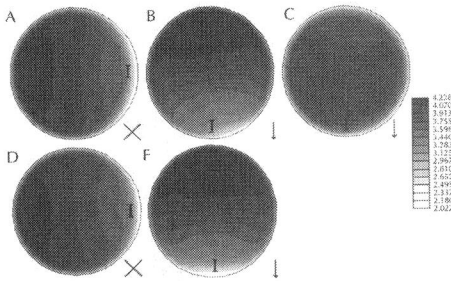

Figure 8: Cross-sectional contour plots of the carrier-phase velocity magnitude at the pipeline stations indicated in Fig. 2. Vectors indicate the

direction of gravity, crosses that the flow is vertical. I designates the inner part of the last bend in the upflow direction from the given cross-section. Hydrate volume fraction: 0.14, mean flow velocity: 3.82 m/s, hydrate particle size: $70\mu m$.

The use of a curvature upstream of the inlet results in a non-uniform velocity profile, which may mimic the flow at the outlet from the centrifugal pump (Cheah et al., 2007). Thus at the inlet to the test section, the velocity maximum was shifted toward one wall (Fig. 8A). Further flow development after the first bend leads to a shift of the velocity maximum toward the top of the first horizontal pipeline segment (Fig. 8B). The contours of velocity magnitude in this cross-section are not symmetrical around a vertical plane through the pipe axis due to the effect of the non-axisymmetrical flow through inlet caused by the pump. The shift in velocity maximum caused by the temperature sensor probe seen in Figs. 6A, B is confirmed by the cross-sectional contour plot presented in Fig. 8C, while the flow asymmetry mentioned at station B is still slightly detectable. From Fig. 8D shows that the maximum in flow velocity again has shifted due to the influence of the second bend, the flow asymmetry is not very clear here and the cross-sectional variation in flow velocity is less. After the third bend, a similar velocity profile is established (Fig. 8F) where the maximum is again associated with the outer part of the bend.

The hydrate-phase volume fraction near the pipeline wall is one of the important parameters which affects the pressure drop via the rheological expression and may well be important for the formation of hydrate plugs in pipelines. Contour plots of the hydrate volume fraction are presented in Figs. 9A–F for the same cross-sectional stations as in Figs. 8A–F again for a flow velocity of 3.82 m/s. Secondary flow vectors for the hydrate phase, i.e. the particle velocity component normal to the axis of the pipe, are also presented in the figures, since the volume-fraction profiles were found to depend on this secondary flow.

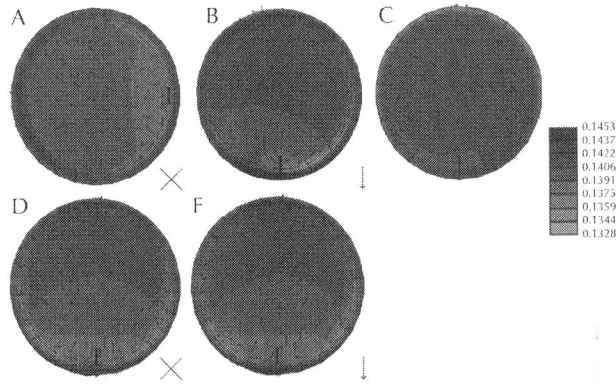

Figure 9: Cross-sectional contour plots of the hydrate volume fraction together with vector plots of the secondary flow normal to the main flow at the pipeline stations indicated in Fig. 2. The primary flow is oriented out of the page. The large vectors illustrate the direction of gravity as in the previous figure. *I* designates the inner part of the last bend in the upflow direction from the given cross-section. Hydrate volume fraction: 0.14, mean flow velocity: 3.82 m/s, hydrate particle size: 70μm.

The hydrate volume fraction at the inlet of experimental pipeline (Fig. 9A) is non-uniform in the near-wall regions due to the influence of the secondary flow structure there. Two Dean-vortices (De=63 902), produced under the effect of inlet curvature mimicking the pump, meet together near the wall at the circumferential position corresponding to the inner part of the curvature, producing a stagnation zone. This zone is associated with a region in which the particle concentration is higher. Another region of particle compaction is located on the opposite wall, and this can be attributed to the particles being centrifuged outward in the curvature. The volume-fraction variation in the bulk of the pipeline is less pronounced than in the near-wall region.

The secondary flow pattern after the first flow loop bend (De=105 969) is non-symmetrical because of the influence of the inlet flow. Two Dean-vortices located there are of different shape and position than the previously mentioned ones, and thus the stagnation zone, associated with compaction of particles, is located closer to the wall (Fig. 9B). The centrifugal effect is also strong in this cross-section producing increased particle density in the top region of the near-wall subsurface. The Dean-vortices break down after the temperature sensor so that the secondary flow is rearranged as shown in Fig. 9C. The region of compacted particles in the stagnation zone seems to be shifted to the right part of near-wall subsurface as we see it. The effect of gravitation has at this point canceled the centrifugal action from the previous cross-section so that there is only one compacted region located in the bottom of the cross-section. Secondary flow structure and volume fraction profiles are similar to ones presented in Fig. 9B because they are produced by a bend with the same Dean number.

For lower mean flow velocities, the secondary flow patterns and the hydrate volume fraction profiles are very similar to the structures presented in Figs. 9A–F. A difference, however, is observed in the horizontal parts of the pipeline, where, on the one hand, the residual effect of centrifugation of particles in the bends becomes less significant and, on the other hand, the settling velocity of the particles more significant relative to the axial velocity, both acting together in causing an increase in the compaction of particles at the bottom of the pipeline due to gravity. The vortices in these cross-sections are also shifted toward the bottom of the pipe due to the interphase momentum exchange.

Decreasing the mean flow velocity below 0.4 m/s causes further decrease in the particle slip velocity (Fig. 12) and in the turbulent mixing of the particles compared with the net force of gravity. This leads to the development of compaction of the hydrate particles in the bottom zones of the horizontal pipeline sections. This effect is visible in Figs. 10A, D which shows the particle concentration in the near-wall subsurface of a horizontal pipeline segment. The contours of hydrate volume fraction in the cross-section of a horizontal pipeline segment are presented in Figs. 10B, F for mean volume fractions of 0.14 and 0.30, respectively. The figures show that the compacted regions are located along the pipeline wall and that the particle concentration in the central part of the cross-section is relatively smaller. The cross-sectional contours of the carrier phase velocity magnitude are presented in Figs. 10C, G for the same hydrate concentrations.

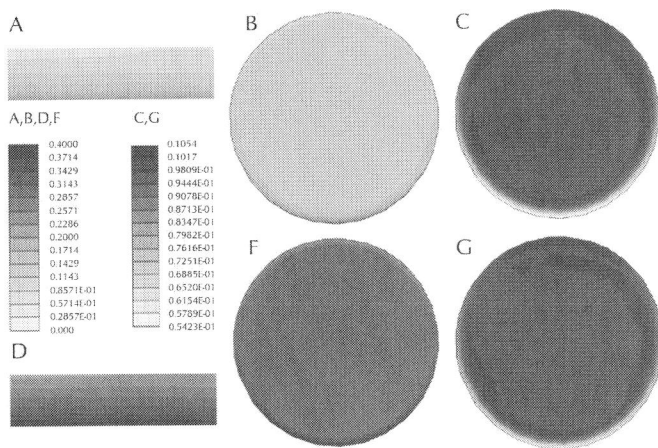

Figure 10: (A, D) Contours of hydrate volume fraction in the near-wall subsurface, (B, F) contours of hydrate volume fraction in the pipeline cross-section marked C in Fig. 2, (C, G) contours of the carrier phase velocity magnitude in the pipeline cross-section marked C inFig. 2. Mean flow velocity: 0.1 m/s, hydrate particle size: 70µm, hydrate volume fractions: 0.14 and 0.30.

Particle Reynolds number, $Re_p \equiv \rho_1 u_r d_p / \mu$ can here be seen as a dimensionless measure of the interphase slip velocity. A contour plot of the relative Reynolds number of the hydrate particles in the midline cross-section is presented in Fig. 11 for a mean flow velocity 3.82 m/s, a hydrate volume fraction of 0.14 and a mean particle size of 70mm The figure shows clearly that maxima in the interphase slip are associated with flow obstructions and with bends as would be expected. Thus the interphase drag term is also maximal in these regions. The information gathered from this type of plot is very interesting and relevant for various aspects of the flow process, such as local rates of particle agglomeration and -breakage and also local rates of mass and heat transfer between the phases. Note, however, that the slip velocity is computed on basis of the time-mean flow, and does not include slip due to turbulent eddies.

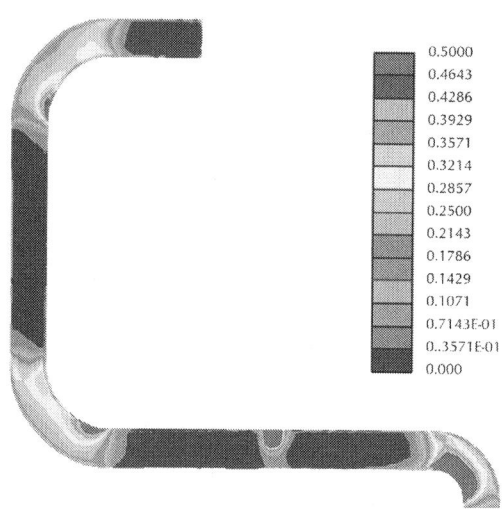

Figure 11: Contour plot of particle Reynolds number in the midline cross-section. Mean flow velocity 3.82 m/s, hydrate volume fraction 14%, mean particle size 70μm.

Fig. 12 presents the cell-average particle Reynolds number as a function of mean flow velocity for hydrate volume fractions of 0.14, 0.22, 0.304 and 0.378 with a mean particle size 70mm The figure shows that the slip velocity increases with increasing pipe velocity as expected. The slip velocity is, as Fig. 11 clearly shows, generated by the presence of the bends and obstructions. This is a very similar situation to inertial separation equipment, in which an acceleration is imparted to the particles generating a slip velocity between particles and fluid to separate the particles by centrifugation or impaction. It is well known that in such equipment the acceleration imparted to the particles is proportional to the square of the velocity of the carrier fluid, for example in cyclones, the centripetal acceleration (negative of the "centrifugal force" in a rotating coordinate system) acting on a particle of mass m swirling in the fluid at radius r is $-mv_\theta^2/r$. On the other hand, the flow force is, assuming Stokes drag law, proportional to the slip velocity, such that at equilibrium the particle's slip velocity is proportional to the square of the tangential velocity of the fluid in the cyclone. In fact, the data shown in Fig. 12 can all be fitted very well with parabolas with vertices at the origin, which is consistent with these notions.

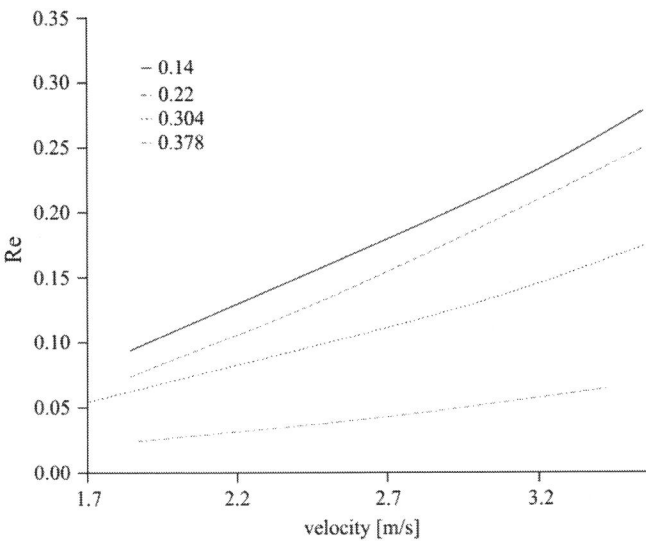

Figure 12: Spatially averaged particle Reynolds number as a function of mean flow velocity for hydrate volume fractions of 0.14, 0.22, 0.304 and 0.378. Mean particle size: 70mm.

The literature on aggregation and breakage of cohesive particles shows that mean particle size in the system is strongly dependent on the shear rate (Hounslow et al., 1988 and Wang et al., 2005) which is related to the turbulent energy dissipation rate (Heath et al., 2006). A contour plot of the dissipation rate is presented in Fig. 13 for the mean flow velocity 3.82 m/s. It can be seen from the plot that the regions with the most intensive dissipation are the inner parts of the bends, shifting to the outside just downstream of the bends, corresponding with the velocity maximum changing location. The energy dissipation rate is also high around the temperature sensor. Thus it is expected that mean particle size would be at the minimum due to the breakage (Herri et al., 1999) while the quiescent regions are aggregation-dominated.

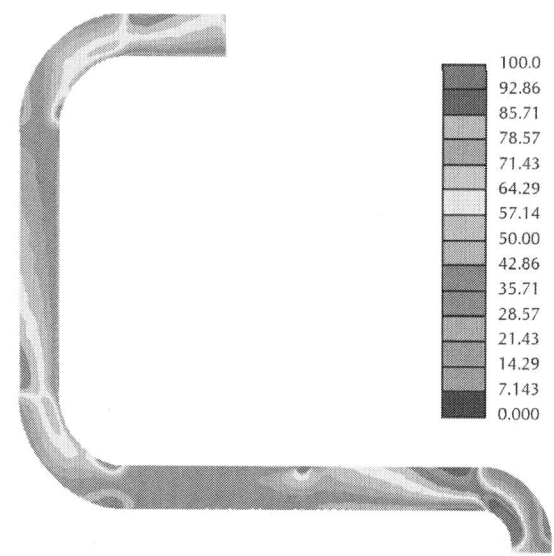

	100.0
	92.86
	85.71
	78.57
	71.43
	64.29
	57.14
	50.00
	42.86
	35.71
	28.57
	21.43
	14.29
	7.143
	0.000

Figure 13: Contour plot of the turbulent kinetic energy dissipation rate on the pipeline wall. Hydrate volume fraction: 0.14, mean flow velocity: 3.82 m/s, hydrate particle size 70mm.

CONCLUSIONS

An experimental and computational study of a low-pressure flow loop loaded with hydrate particles has been performed. It was found during experiments that the presence of hydrate phase increases the total system pressure loss resulting in an increase in the pumping work necessary to establish hydrate slurry recirculation. Isokinetic sampling enabled the determination of the mean hydrate particle size for three different concentrations. An increase in the particle size was found with increasing volume fraction, which can be considered to be a consequence of their growth and flocculation. Local increases in hydrate phase-concentration by the lower wall were observed in the horizontal pipeline regions for the lower mean flow velocities of 0.1–0.4 m/s. The experimental rig was plugged with hydrate phase at the relatively high volume fraction of 40%, as the pressure drop and flowrates generated by the pump were high enough to prevent possible pipeline blockage at lower hydrate concentrations in the system.

A CFD-model was built for the reproduction of experimental results and providing the insight into the flow characteristics. Two simplifying assumptions were made in the model. The first assumption was that of a uniform and constant hydrate particle size of 70μm. The second assumption was that the apparent viscosity of the hydrate slurry follows a dependence obtained from experimental measurements of the pipeline pressure drop. The model was validated with experimental pressure gradients and a discrepancy of less than 8% was found. Dependencies of the velocity and volume fraction profiles on the model geometry (bends and fixtures and the orientation of the pipeline) as well as on the average flow velocity and over-all hydrate volume fraction were found.

The process of local particle densification at a variety of mean velocities and for different average hydrate concentrations was studied computationally. Turbulent energy dissipation rate and particle Reynolds number maxima were charted and shown to be associated with local flow disturbances produced by the obstacles.

ACKNOWLEDGMENTS

The authors acknowledge StatoilHydro, Chevron ETC, SINTEF Petroleum Research and the Norwegian Research Council for funding through the HYADES project and for permission to publish the data shown in this article.

REFERENCES

1.   Balakin, B.V., Hoffmann, A.C., Kosinski, P., 2010. Population balance model for nucleation, growth, aggregation and breakage of hydrate particles in turbulent flow. AIChE Journal, doi:10.1002/aic.

2. Balakin, B.V., Pedersen, H., Kilinc, Z., Hoffmann, A.C., Kosinski, P., Hoiland, S., 2010. Turbulent flow of Freon R11 hydrate slurry. Journal of Petroleum Science and Engineering 70, 177–182.

3. Berge, L.I., Gjertsen, L.H., Lysne, D., 1998. Measured permeability and porosity of R11 hydrate plugs. Chemical Engineering Science 53, 1631–1638.

4. Cheah, K.W., Lee, T.S., Winoto, S.H., Zhao, Z.M., 2007. Numerical flow simulation in a centrifugal pump at design and off-design conditions. International Journal of Rotating Machinery, 1–8.

5. Crowe, C., Sommerfeld, M., Tsuji, Y., 1998. Multiphase Flows with Droplets and Particles. CRC Press, Boca Raton.

6. Darbouret, M., Cournil, M., Herri, J.-M., 2005. Rheological study of TBAB hydrate slurries as secondary two-phase refrigerants. International Journal of Refrigeration 28, 663–671.

7. Darbouret, M., Le Ba, H., Cameirao, A., Herri, J.-M., Peytavy, J.-L., Glenat, P., 2008. Lab scale and pilot scale comparison of crystallization of hydrate slurries from a water-in-oil emulsion using chord length measurements. In: Proceedings of the 6th International Conference on Gas Hydrates (ICGH 2008), Vancouver, British Columbia, Canada, July 6–10.

8. Darby, R., 2001. Correlate pressure drops through fittings. Chemical Engineering 108, 127–130. de Nevers, N., 1991. Fluid Mechanics for Chemical Engineers, second ed. McGrawHill Science/Engineering/Math ISBN 0-201-01497-1.

9. Ergun, S., 1952. Fluid flow through packed columns. Chemical Engineering Progress 48, 89–94.

10. Ferziger, J.H., Peric, M., 2001. Computational Methods for Fluid Dynamics, third ed. Springer, Heidelberg.

11. Greaves, D., Boxall, J., Mulligan, J., Sloan, E.D., Koh, C.A., 2008. Hydrate formation from high water content-crude oil emulsions. Chemical Engineering Science 63, 4570–4579.

12. Gidaspow, D., 1994. Multiphase Flow and Fluidization: Continuum and Kinetic Theory Descriptions. Academic Press, Boston.

13. Hashemi, S., Macchi, A., Servio, P., 2009. Gas–liquid mass transfer in a slurry bubble column operated at gas hydrate forming conditions. Chemical Engineering Science 64, 5907–5912.

14. Heath, A.R., Bahri, P., Fawell, P., Farrow, J., 2006. Polymer flocculation of calcite: experimental results from turbulent pipe flow. AIChE Journal 52, 1284–1293.

15. Herri, J.M., Pic, J.S., Gruy, F., Cournil, M., 1999. Methane hydrate crystallization mechanism from in-situ particle sizing. AIChE Journal 45, 590–602.

16. Holbeach, J.W., Davidson, M.R., 2009. An Eulerian–Eulerian model for the dispersion of a suspension of microscopic particles injected into a quiescent liquid. Engineering Applications of Computational Fluid Mechanics 3, 84–97.

17. Hounslow, M.J., Ryall, R.L., Marshall, V., 1988. A discretized population balance for nucleation, growth, and aggregation. AIChE Journal 34, 1821–1832.

18. Huang, H., Fake, D.M., Calkins, W.H., Klein, M.T., 1994. A novel laboratory scale reactor system for studying fuel processes from the initial stages. 2. Detailed energy and momentum balances. Energy and Fuels 8, 1310–1315. J

19. assim, E., Abdi, M., Muzychka, Y., 2008. A CFD-based model to locate flow restriction induced hydrate deposition in pipelines. Offshore Technology Conference.

20. Methodology, 2006. STAR-CD Version 4.02. CD-Adapco, 200 Shepherds Bush Rd., London. Patankar, N.A., Joseph, D., 2001. Modeling and numerical simulation of particulate flows by the Eulerian–Lagrangian approach. International Journal of Multiphase Flow 27, 1659–1684.

21. Pfleger, D., Gomes, S., Gilbert, N., Wagner, H.-G., 1999. Hydrodynamic simulations of laboratory scale bubble columns fundamental studies of the Eulerian– Eulerian

modelling approach. Chemical Engineering Science 54, 5091–5099.

22. Sean, W.-Y., Sato, T., Yamasaki, A., Kiyono, F., 2007. CFD and experimental study of methane hydrate dissociation. AIChE Journal 53, 262–274.

23. Sinquin, A., Palermo, T., Peysson, Y., 2004. Rheological and flow properties of gas hydrate suspensions. Oil and Gas Science and Technology 28, 37–50.

24. Sloan, E.D., Koh, C.A., 2008. Clathrate Hydrates of Natural Gases, third ed. CRC Press, 6000 Broken Sound Parkway NW, Suite 300, Boca Raton, FL, USA.

25. Viessman, W., Hammer, M.J., 2004. Water Supply and Pollution Control, seventh ed. Prentice Hall, Inc., Upper Saddle River, NJ, USA. Wang, L., Marchisio, D.L., Vigil, R.D., Fox, R.O., 2005. CFD simulation of aggregation and breakage processes in laminar Taylor–Couette flow. Journal of Colloid and Interface Science 282, 380–396.

26. Wang, W., Fan, S., Liang, D., Yang, X., 2008. Experimental study on flow characters of CH_3CCl_2F hydrate slurry. International Journal of Refrigeration 31, 371–378.

A Transient Outflow Model for Pipeline Puncture

Adeyemi Oke[a], Haroun Mahgerefteh[a], Ioannis Economou[b], and Yuri Rykov[c]

[a]Department of Chemical Engineering, University College London, Torrington Place London WC1E 7JE, UK

[b]NRCPS "Demokritos" , Institute of Physical Chemistry, GR-153 10 Aghia Paraskevi Attikis, Greece

[c]Keldysh Institute of Applied Mathematics, Miusskaya sq. 4, 125047 Moscow, Russia

ABSTRACT

This paper describes the development and validation of a highly efficient robust numerical simulation based on the method of characteristics for predicting release rates following the puncture of pipelines containing high pressure hydrocarbon mixtures. The model accounts for real fluid behaviour, radial and axial flow as well as the locality of puncture relative to the length of the pipeline. It is applicable to both isolated and un-isolated flows where pumping at the high-pressure end continues despite puncture. The results of the application of the model to the hypothetical puncture of a 16 km, 0.42 m dia. pressurised pipeline containing a condensable hydrocarbon mixture are presented as a case example. Simulated fluid velocity and pressure profiles are used to provide a pictorial timeline representation of the important post-puncture fluid dynamics phenomena within the pipeline, which ultimately govern the discharge process. These results indicate that the conventional outflow models treating the pipeline as a vessel discharging through an orifice are inappropriate especially during the early stages of depressurisation. Good agreement is obtained between the results of the simulation when compared to experimental data from the puncture of 100 m long LPG pipeline.

INTRODUCTION

Large amounts of highly flammable pressurised hydrocarbons are frequently transported in long pipelines across the world. These represent a very serious safety hazard, which in the event of pipeline rupture can cause significant fatalities and damage to the environment. The prediction of the ensuing release rate and its variation with time are the two critical pieces of information required in assessing and quantifying the consequences associated with such failures. In the offshore industry for example, such data may dictate the survival time of the mechanical integrity of the production platform. Indeed, during the Piper Alpha tragedy (Cullen, 1990), the heat intensity following the rupture and ignition of the main gas export pipeline resulted in the eventual collapse of the platform into the seabed and the loss of 167 lives. There are numerous other incidents involving pipeline rupture, which have resulted in significant fatalities (Bond, 2002; Fletcher, 2001a). The impact on the environment following pipeline failure can also be equally devastating. An average of over 6.3 million gallons of oil and other hazardous liquids are reported released from pipelines each year, more than half the amount released from the Exxon Valdez disaster (Bond, 2002). Record fines exceeding hundreds of millions of dollars are now being imposed on pipeline operators causing damage to the environment (Fletcher, 2001a; True, 2001).

Following considerable public concern over pipeline safety, US Senate Committee on Commerce, Science, and Transportation is proposing 'The Pipeline Safety Improvement Act (S2438)' requiring stringent pipeline safety requirements in high-consequence areas (Fletcher 2001a and Fletcher 2001b;Barlas, 1999). Those pipeline operators who do not have the necessary expertise in modelling the consequences of pipeline failure and hence demonstrate the level of the hazard will face major problems in the event of the above legislation being passed through the Senate.

The mathematical problem to be solved for simulating outflow comprises the equations of mass, momentum and energy conservation for turbulent flow, which is, in general, partly single-phase, and partly two-phase flow (Zucrow & Hoffman, 1976). Due to the fact that these conservation equations contain terms that cannot be resolved analytically, they can only be solved using a numerical technique (Flatt, 1986). The modelling also involves cross-sectionally averaged quantities, so that it is one-dimensional in space along the pipeline. Dissipative terms involving heat transfer and friction may be modelled using suitable correlations. Due to the large pressure drops involved, the model is stiff, and this is one cause of the inefficiency of current numerical procedures. Computation run times of few weeks for simulating the complete evacuation of long (>100km) pipelines containing pressurised hydrocarbons using even relatively fast (e.g. 1.7 GHz Pentium IV) computers are not uncommon (Mahgerefteh, Saha, & Economou, 1999).

In recent publications (Mahgerefteh, Saha, & Economou, 1997, Mahgerefteh et al., 1999, Mahgerefteh, Saha, & Economou, 2000), we described the development of a relatively efficient numerical simulation based on the method of characteristics for pipeline rupture. As well as accurately predicting outflow when compared to real data, the simulation was extensively used to simulate the dynamic behaviour of emergency shutdown valves following emergency isolation. The prevailing fluid dynamic profile within the pipeline during the release process was shown to critically influence the efficacy and mechanical integrity of different types of emergency shutdown valves used for isolating the flow.

Despite the success of our model in simulating real data, its results were limited to full bore rupture failure in which the pipeline is effectively split into two across its circumference. Although such type of pipeline failure is the most catastrophic, research sponsored by the Swedish Nuclear Power Inspectorate (Lydell, 2000) has shown that of the 3751 pipe failure events recorded between 1994 and 1999, more than two thirds were in the form of leaks or punctures. For these failures, flow in the proximity of the puncture will be both in the axial and the radial directions.

It is worth noting that in recent years, a number of models with varying degrees of sophistication (e.g. isothermal or isentropic flow) for pipeline puncture have been reported (see for example Woodward & Mudan, 1991). These in principle treat the pipeline as a vessel discharging through an orifice, thereby ignoring the effects of the ensuing pressure and fluid flow transients within the pipeline. Others (see for example Montiel, Vilchez, Casal, & Arnaldos, 1998; Young-Do and Bum, 2003) are based on steady state adiabatic flow assumptions ignoring real fluid behaviour. Rapid depressurisation results in propagation of expansion waves from the rupture point to the low-pressure end of the pipeline. In the case of full bore rupture, the speed at which these waves propagate critically influence the discharge rate (Zucrow & Hoffman, 1976; Mahgerefteh et al., 1999). The significance of such phenomena on the release process during pipeline puncture is not known. In addition, the simple models are incapable of simulating the highly plausible failure scenario involving un-isolated release where pumping of the pressurised inventory continues despite puncture.

Based on the solution of the conservation equations using a finite difference method of analysis, Chen, Richardson, & Saville 1992, Chen, Richardson, & Saville 1993, Chen, Richardson, & Saville 1995a and Chen, Richardson, & Saville 1995b pipeline rupture model is in principle capable of simulating pipeline puncture. However despite its mathematical rigour and good performance in simulating FBR, it is based on uni-axial flow thus applicable to modelling pipeline fluid flow well away from non-axisymmetric failures.

Bendiksen, Malnes, Moe, and Nul (1991) pipeline rupture model, OLGA incorporates a heterogeneous equilibrium model which heavily relies on empirically obtained data. This makes OLGA prone to errors, introducing gross inaccuracies in its predictions when compared to real data (Shoup, Xiao, & Romma, 1998).

Zhou, Lea, Bilo, and Maddison (1997) model incorporates geometric and physical complexities that may exist in the pipe system including puncture located at any point and orientation along the pipeline. Although the model gives exhaustive description of the fluid thermophysical properties including 3-D simulation of the escaped gas trajectory, it is nonetheless based on steady state flow assumption involving isentropic release at the puncture plane.

This paper describes the development of a highly efficient robust numerical simulation for predicting outflow following the puncture of pressurised pipelines. The model accounts for real fluid behaviour, radial and axial flow at the puncture plane as well as the location and orientation of puncture along the pipeline. The specially developed solution technique based on pressure–enthalpy flash results in a significant reduction in computing run times. This development addresses a major difficulty synonymous with the numerical simulation of pipeline ruptures.

The model is applied to the hypothetical puncture of a pressurised pipeline containing a condensable hydrocarbon mixture. Both isolated and un-isolated failure scenarios where feeding continues despite puncture are simulated. By simulating the transient fluid velocity and pressure profiles, we provide a detailed insight into some intriguing but nevertheless important fluid flow phenomena occurring within the pipeline that govern the discharge process following puncture. Finally, the model is validated by comparison with Isle of Grain (Richardson & Saville, 1996) experimental data for puncture of a 100 m LPG pipeline.

THEORY

In the case of unsteady generalised 1-D flow, assuming thermodynamic and phase equilibrium between the constituent phases, the respective continuity, momentum, and energy conservation equations for an element of a fluid in a rigidly clamped pipeline are given by Versteeg and Malalasekera (1995):

$$\frac{d\rho}{dt} + \rho \frac{\partial u}{\partial x} = 0 \tag{1}$$

$$\rho \frac{\partial u}{\partial t} + \rho u \frac{\partial u}{\partial x} + \frac{\partial P}{\partial x} = \alpha, \tag{2}$$

$$\rho \frac{dh}{dt} - \frac{dP}{dt} - (q_h - u\beta_y) = 0 \tag{3}$$

where ρ, u, P and h are the density, velocity, pressure and specific enthalpy of the fluid as a function of time, t and distance, x, respectively.

The heat transfer across the pipe wall to the fluid is given by

$$q_h = \frac{4}{D} U_h(T_{amb} - T_f) \tag{4}$$

where U_h, is the overall heat transfer coefficient, with T_{amb} and T_f denoting the ambient and the fluid temperatures, respectively.

βy, is the friction force term, in turn given by

$$\beta_y = -2\frac{f_w}{D} \rho u |u| \tag{5}$$

where, f_w is the fanning friction factor and D, the pipeline diameter.

Also,

$$\alpha = -\left(\frac{2f_w \rho u|u|}{D} + \rho g \sin \theta\right)$$

(6)

where ϑ, is the angle of inclination of the pipeline to the horizontal plane.

Based on, and , given a function $f(x,t)=\rho,u,P$ or h, we have

$$\frac{df}{dt} = \frac{\partial f}{\partial t} + u \frac{\partial f}{\partial x}$$

(7)

With the aid of suitable thermodynamic transformations, the total derivative of density with respect to time in Eq. (1) can be obtained as

$$\frac{d\rho}{dt} = \frac{1}{a^2}\left[\frac{dP}{dt}\left(1 + \frac{\varphi}{\rho T}\right) - \frac{\varphi}{T}\frac{dh}{dt}\right]$$

(8)

The speed of sound, a and the thermodynamic function, φ for real multi-component single-phase fluids can be derived as

$$a^2 = \frac{\gamma}{k\rho},$$

(9)

$$\varphi = \left(\frac{\partial P}{\partial s}\right)_\rho = \frac{a^2 \rho \xi T}{C_P \gamma - Ta^2\xi^2}$$

(10)

where, γ is the ratio of specific heats, T, k and ξ, respectively are the fluid temperature (K), isothermal and isobaric coefficients of volumetric expansion. C_p, is the specific heat capacity at constant pressure.

For two-phase flows, the analytical determination of γ and Cp becomes complex (Mahgerefteh et al., 1999). Hence the parameters, a and φ are evaluated numerically at a given temperature and pressure. Accordingly

$$a^2 = \left(\frac{\Delta P}{\Delta \rho}\right)_s$$

(11)

$$\varphi = \rho^2 \left(\frac{\Delta T}{\Delta \rho}\right)_s$$

(12)

The flow dependent wall friction force term, $_y$ in and is obtained using the Moody approximation (Massey, 1983) to the Colebrooke equation. Gas and liquid viscosities required for the calculation of $_y$ are obtained according to the Ely and Hanley scheme for non-polar gaseous mixtures, and the Dymond and Assael scheme for liquid mixtures (Massey, 1983).

We employ the Peng–Robinson EoS (Peng & Robinson, 1976) for obtaining the appropriate thermodynamic and phase equilibrium data. This equation has been shown to be particularly applicable to high-pressure hydrocarbon mixtures (see for example de Reuck, Craven, & Elhassan, 1996; Assael, Martin Trusler, & Tsolakis, 1996). However care should be taken in the accuracy of its predictions near the critical region. The number and the appropriate fluid phase(s) present at any given temperature and pressure on the other hand are determined using the stability test based on the Gibbs tangent plane criterion developed by Michelsen (Michelsen 1982a, Michelsen 1982b and Michelsen 1987). For unstable systems, the same technique also provides the composition of a new phase, which can be split off to decrease the Gibbs energy of the mixture.

Pseudo-fluid properties for mixtures are calculated from the pure liquid and gas properties obtained from the EoS.

Based on Eq. (8), the continuity equation can be reformulated as

$$[\rho T + \varphi]\frac{\mathrm{d}P}{\mathrm{d}t} - \rho\varphi\frac{\mathrm{d}h}{\mathrm{d}t} + \rho^2 a^2 T\frac{\partial u}{\partial x} = 0.$$

(13)

Eqs. (13), (2), and (3), respectively, represent the continuity, momentum and energy equations on which the puncture model is based. It can be seen that none of the derivatives in these equations are in terms of the fluid density. As a result the corresponding solution of the system of equations may be expressed in terms of fluid pressure, enthalpy and velocity.

The conservation equations may be solved using a variety of numerical methods with varying degrees of success in terms of accuracy and computational run time. These include finite difference method (FDM) (Chen, Richardson, & Saville 1993,

Chen, Richardson, & Saville 1995a and Chen, Richardson, & Saville 1995b; Bendiksen et al., 1991), a finite element method (FEM) (Lang, 1991, Bisgaard, Sorensen, & Spangenberg, 1987) and the method of characteristics (MOC) (Chen et al., 1992; Flatt, 1986; Olorunmaiye & Imide, 1993). Of these, MOC is considered as the numerical benchmark, having been shown (Mahgerefteh, Saha, & Economou 1997 and Mahgerefteh, Saha, & Economou 1999) to simulate outflow following full-bore rupture of long pipelines with a good degree of accuracy.

Using MOC, the conservation Equations (13) and may be replaced with three compatibility equations along their corresponding characteristics lines (Zucrow & Hoffman, 1976). The characteristic lines in essence govern the speed at which expansion waves propagate from the rupture/puncture point to the low and high pressure ends of the pipeline (positive and negative Mach lines), while the path line dictates the rate of flow through any given point along the pipeline. The corresponding compatibility equations may be derived from first principles following Zucrow and Hoffman (1976) and are given as

$$\rho d_0 h - d_0 P = \psi d_0 t,$$

along the path line characteristic;

$$\frac{d_0 t}{d_0 x} = \frac{1}{u},$$

$$\rho a d_+ u + d_+ P = \left[a\alpha + \frac{\varphi\psi}{\rho T} \right] d_+ t,$$

$$(14)$$

along the positive Mach line characteristic;

$$\frac{d_+ t}{d_+ x} = \frac{1}{u + a},$$

$$\rho a d_- u - d_- P = \left[a\alpha - \frac{\varphi\psi}{\rho T} \right] d_- t,$$

$$(15)$$

along the negative Mach line characteristic;

$$\frac{\mathrm{d}\_t}{\mathrm{d}\_x} = \frac{1}{u-a},$$

(16)

where the non-isentropic term ψ, incorporating heat transfer and frictional effects, is given by

$$\psi = q_h - u\beta_y$$

(17)

Fig. 1 shows a schematic representation of the three characteristics lines plotted on the time (t) and distance (x) axes along the pipeline. The values of P, h, ρ, u, and a, as a function of time and distance along the pipeline (say points, p, o and n) are obtained by the inverse marching method of characteristics (Chen et al., 1992). This involves dividing the pipeline into a large number of distance (Δx) and time elements (Δt) and expressing the compatibility equations in finite difference form. These are in turn solved at the intersection of the linear characteristics lines with the spatial axis (i.e. points p,o, n and the intersection of the characteristics lines, point j) using iteration and linear interpolation in conjunction with P–h (pressure–enthalpy) flash calculations using the Peng–Robinson equation of state. The model throughout its formulation employs linear characteristics and linear interpolations as opposed to curved characteristics (Mahgerefteh et al., 1999). The former has been found to lead to greater numerical stability especially around the phase transition boundary.

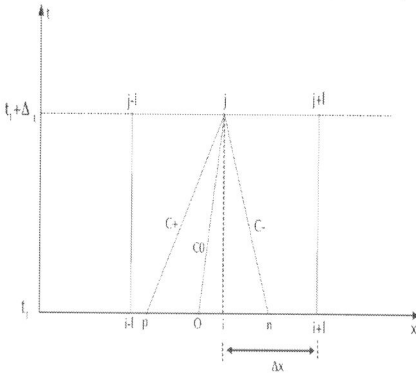

Figure 1: Representation of Path and Mach line characteristics at a grid point.

First-order finite difference approximations (see Eqs. (18)–(20)) to the compatibility equations are employed for the first iteration step, while for subsequent steps, second-order approximations (see Eqs. (18)–(20)) are utilised.

The first-order approximations are given by

$$(\rho)_0(h_j - h_0) - (P_j - P_0) = \psi_0(t_j - t_0). \tag{18}$$

Positive Mach line compatibility:

$$(\rho a)_p(u_j - u_p) + (P_j - P_p)$$

$$= \left(a\alpha + \frac{\varphi\psi}{\rho T} \right)_p (t_j - t_p). \tag{19}$$

Negative Mach line compatibility:

$$(\rho a)_n(u_j - u_n) - (P_j - P_n)$$

$$= \left(a\alpha - \frac{\varphi\psi}{\rho T} \right)_n (t_j - t_n). \tag{20}$$

While the second-order approximations are given by

Path line compatibility,

$$\tfrac{1}{2}[(\rho)_0 + (\rho)_j](h_j - h_0) - (P_j - P_0)$$

$$= \tfrac{1}{2}[\psi_0 + \psi_j](t_j - t_0). \tag{21}$$

Positive Mach line compatibility,

$$\frac{1}{2}[(\rho a)_p + (\rho a)_j](u_j - u_p) + (P_j - P_p)$$

$$= \frac{1}{2}\left[\left(a\alpha + \frac{\varphi\psi}{\rho T} \right)_p + \left(a\alpha + \frac{\varphi\psi}{\rho T} \right)_j \right](t_j - t_p). \tag{22}$$

Negative Mach line compatibility,

$$\frac{1}{2}[(\rho a)_n + (\rho a)_j](u_j - u_n) - (P_j - P_n)$$

$$= \frac{1}{2}\left[\left(a\alpha - \frac{\varphi\psi}{\rho T} \right)_n + \left(a\alpha - \frac{\varphi\psi}{\rho T} \right)_j \right](t_j - t_n). \tag{23}$$

The complete solution for the entire length of the pipeline involves the introduction of boundary conditions at the pertinent nodes situated at the inlet and exit points of the fluid in the pipeline.

BOUNDARY CONDITIONS

The failure scenario simulated in this study involves the puncture of a pipeline at any point along its length followed by instantaneous closure of an emergency isolation valve. Both isolated and un-isolated flows are to be modelled. The former is based on cessation of inlet feed upon puncture. In the case of un-isolated flow on the other hand, feed inlet is assumed to terminate after a prescribed period of time following puncture.

The above modelling requires the imposition of appropriate boundary conditions on the compatibility Eqs. (14)–(16) at the pertinent nodes along the pipeline. These enable closure of these equations with their solutions establishing the fluid thermophysical properties in the time and space grids. Fig. 2 is a schematic representation of fluid flow analysis following puncture and emergency isolation. Pipeline 1 and 2 refer to upstream and downstream of the puncture, respectively. The various boundary conditions, B_1–B_6 indicated in the diagram are placed at the following locations and conditions along the pipeline:

- Reservoir/pump inlet, B_1.
- Cessation of pumping at inlet, B_2.
- Downstream end of pipeline Section 1, B_3.
- Common junction at puncture point, B_4.
- Upstream end of pipeline section 2, B_5.
- Intact end, B_6.

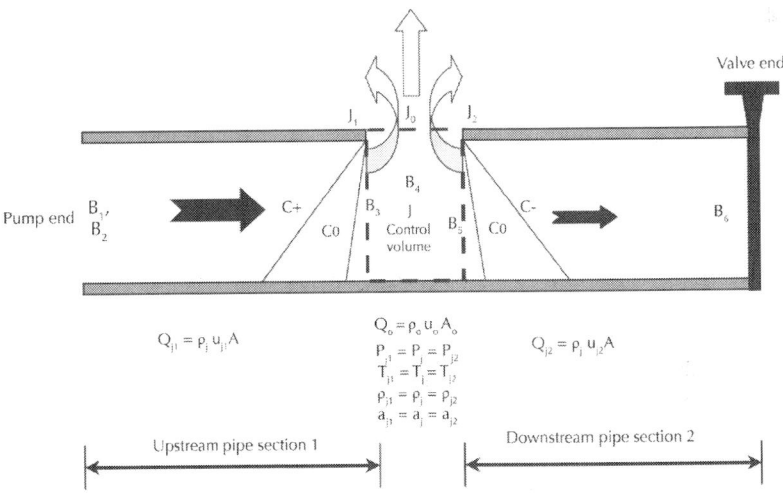

Figure 2: Schematic representations of fluid flow and puncture plane analysis following line puncture.

For the downstream boundary conditions (B_3 and B_6), only the positive Mach line (C_+) and path line (C_0) characteristics hold. For upstream boundary conditions (B_1, B_2 and B_5) on the other hand, only the negative Mach line (C_-) and path line characteristics are applicable.

Reservoir/Pump Inlet, $\boldsymbol{B}_1$

For an infinite reservoir such as discharge from a wellhead, the pipeline inlet pressure is assumed to remain constant. Thus, the pressure, Pj at the solution point, j for a given time duration, tj is fixed. Consequently, the remaining variables that are solved for at the solution point are the fluid specific enthalpy, hj and velocity, uj. These are in turn obtained from the iterative solution Eqs. (21) and (23) in conjunction with the equation of state.

In the case of a pump operating at pipeline inlet on the other hand, an equation representing the pump*discharge curve* or *characteristics* is required. In this study, the equation for centrifugal pumps given byWylie and Streeter (1993) is employed:

$$P_j = P_{SH} + u_j(K_{p1} + K_{p2}u_j), \quad (24)$$

where P_{SH} is the pump shut-off head, and K_{p1} and K_{p2} are constants whose magnitude depend on the type of pump used. P_j and u_j are respectively the pump discharge pressure and velocity. Eqs. (21), (23), and (24) solved iteratively in conjunction with the equation of state to obtain the solution point pressure, P_j, enthalpy, h_j and velocity, u_j. Eq. (24) is substituted into Eq. (23) to obtain a quadratic expression the positive root for which corresponds to, u_j.

Cessation of flow at pipeline inlet, B_2

In this case, the values of $P, h, a,$ and T at the solution point (j) are unknown and must be solved for iteratively in conjunction with Eqs. (21) and (23). The speed of sound, a fluid temperature, T and other pertinent thermodynamic fluid properties are obtained by a pressure–enthalpy (P–h) flash calculation at the solution point. Cessation of feed imposes a constraint of zero flow at the pipeline inlet ($uj=0$).

Downstream end of Pipeline Section 1, B_3

None of the flow variables at the downstream end of pipeline section 1 are known. However, their values determine P, ρ and h at the orifice junction (B_4) and upstream of the second pipeline section (B_5). The boundary conditions at the orifice junction force the solution point pressures, Pj_1 (pipeline section 1 downstream pressure), Pj_2 (pipeline section 2 upstream pressure) and the junction pressure, Pj to be equal. This is a consequence of the application of transient flow boundary conditions to the common junction at the puncture point (Wylie & Streeter, 1993). These stipulate that the pressure, Pj at the junction should be equal to the upstream and downstream pressures of all pipe and non-pipe elements that emanate or terminate from it.

The junction pressure, Pj and enthalpy, hj are obtained using an iterative technique based on root bracketing coupled with the Brent method (Press, Teukolsky, Vetterling, & Flannery, 1992) in conjunction with momentum, energy and mass balances conducted across a control volume. The procedure is described later.

Common Junction at Puncture Point, $\boldsymbol{B}_4$

At the common junction, $Pj=Pj_1=Pj_2$. We assume that within the small discretisation time step, Δt the flow through the puncture plane and the control volume that bound the puncture (see Fig. 2) is isentropic. For relatively small time steps, contact area, and high flow rates, as is commonly the case at the junction, this approximation is considered valid. Considering the control volume around the junction point depicted in Fig. 2, the conservation equations for axial and radial flow in terms of Cartesian coordinates can be derived as (Versteeg & Malalasekera, 1995)

Continuity:

$$\frac{d\rho}{dt} + \rho\frac{\partial u}{\partial x} + \rho\frac{\partial v}{\partial y} = 0.$$

(25)

Momentum in the x-direction:

$$\rho\frac{du}{dt} = -\frac{\partial P}{\partial x} - \rho g \sin\theta + \beta_y.$$

(26)

Momentum in the y-direction:

$$\rho\frac{dv}{dt} = -\frac{\partial P}{\partial y} - \rho g \cos\phi + \beta_x.$$

(27)

Energy:

$$\rho\frac{\partial H}{\partial t} + \rho g\frac{\partial z}{\partial t} + \rho u\left(\frac{\partial H}{\partial x} + g\frac{\partial z}{\partial x}\right)$$

$$+ \rho v\left(\frac{\partial H}{\partial y} + g\frac{\partial z}{\partial y}\right) - \frac{\partial P}{\partial t} = q_h.$$

(28)

where

$$H = h + \frac{u^2}{2} + \frac{v^2}{2},$$

(29)

$$\beta_x = -2\frac{f_{pc}}{D_{pc}}\rho v|v|.$$

(30)

Also by definition, for any function $f(x,y,t), f=P, ,H,h,u$, we have

$$\frac{df}{dt} = \frac{\partial f}{\partial t} + u\frac{\partial f}{\partial x} + v\frac{\partial f}{\partial y},$$

(31)

where, u and v are the velocity components in the x and y directions, respectively.

z and ϑ on the other hand are respectively the pipeline elevation and angle of elevation from the horizontal plane. f_{pc} is the Fanning friction factor for flow across the puncture with D_{pc}, denoting the equivalent diameter of the puncture. φ, is the angle that the velocity component, v makes with the vertical plane that divides the pipeline into symmetrical halves; see Fig. 3. Hence

$$\frac{\partial z}{\partial x} = \sin\theta,$$

(32)

$$\frac{\partial z}{\partial y} = \cos\phi.$$

(33)

Also, since the inclination is time invariant, then

$$\frac{\partial z}{\partial t} = 0.$$

(34)

Substituting Eqs. (26), (27), (29),(32), (33), and (34) into Eq. (28) and assuming isentropic flow (i.e., $\beta_x=\beta_y=0$ and $q_h=0$), the energy Eq. (28) reduces to

$$\rho\frac{dh}{dt} - \frac{dP}{dt} = 0.$$

(35)

Multiplying Eq. (25) by V/ρ, where V is the control volume and noting that

$$V = \int_{z_{orth1}}^{z_{orth2}} \int_{y_1}^{y_2} \int_{x_1}^{x_2} \partial x \, \partial y \, \partial z_{orth},$$

(36)

$$A_{yz} = \int_{z_{orth1}}^{z_{orth2}} \int_{y_1}^{y_2} \partial y \, \partial z_{orth},$$

(37)

$$A_{xz} = \int_{z_{orth1}}^{z_{orth2}} \int_{x_1}^{x_2} \partial x \, \partial z_{orth},$$

(38)

where, A_{yz} and A_{xz} are the pipeline and orifice cross sectional areas, respectively. z_{orth} is the plane orthogonal to the x–y plane such that x–y and z_{orth} form a right handed system. Also, x_1, x_2, y_1, y_2, z_{orth1}, and z_{orth2} are the spatial boundaries of the control volume along the x, y and z_{orth} axes.

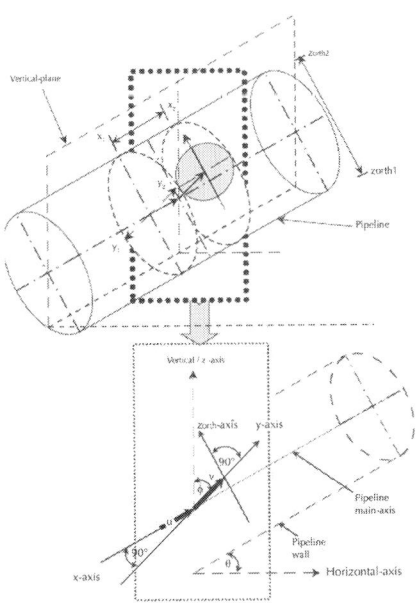

Figure 3: A schematic representation of flow and space variables following line puncture.

Eq. (25) becomes

$$\frac{V}{\rho}\frac{d\rho}{dt} + A_{yz}\int_{u_1}^{u_2} du + A_{xz}\int_{v_1}^{v_2} dv = 0.$$

(39)

Furthermore, by substituting for the total derivative of pressure in Eq. (8) using Eq. (35) and inserting the final expression obtained into Eq. (39) we have

$$\frac{V}{\rho}\frac{1}{a^2}\left[\rho\frac{dh}{dt}\right] + A_{yz}\int_{u_1}^{u_2} du + A_{xz}\int_{v_1}^{v_2} dv = 0.$$

(40)

Eq. (40) can be expressed in integral form as

$$\int_{h_1}^{h_2}\frac{V}{a^2} dh + A_{yz}\int_{t=t_1}^{t=t_2}\int_{u_1}^{u_2} du\, dt$$

$$+ A_{xz}\int_{t=t_1}^{t=t_2}\int_{v_1}^{v_2} dv\, dt = 0.$$

(41)

The integrals in the above expression can be solved numerically using the trapezoidal rule. For a small time step, Δt, Eq. (41) can be expressed as

$$\left(\frac{V}{a^2}\right)_{ave}(h_{j|_{t=t_2}} - h_{j|_{t=t_1}})$$

$$+ [A_{yz}(u_{j|_{x=x_2}} - u_{j|_{x=x_1}})]_{ave}$$

$$+ A_{xz}(v_{|_{y=y_{pc}}})_{ave}]\Delta t = 0,$$

(42)

where the subscript, ave represents the average of the value in the brackets between time, $t=t_1$ and $t=t_2$, such that $t_2-t_1=\Delta t$. x_2 and x_1, respectively, represent the upper and lower spatial boundaries along the x-axis of the control volume, V. There is no flow into the control volume along the y-axis, hence $v_{|y=y1}=0$. Also $u_{j|x=x2}$ and $u_{j|x=x1}$ are equal to u_{j2} and u_{j1}, respectively.

The solution point flow velocities, uj_1 and uj_2 are respectively determined from the application of the appropriate boundary conditions, B_3 given above, and B_5 described below, respectively.

The time step, Δt is chosen subject to the Courant stability criterion (Courant, Friedrichs, & Lewy 1926).

Upstream end of Pipeline Section 2, B_5

The boundary condition imposed at the inlet to the second pipeline section is such that apart from the fluid velocity, uj_2 all the values of the flow variables (P, h, T, ρ, and a) at the common junction of the two pipe sections and the puncture are equal. uj_2, is therefore obtained from the solution of equation (23).

Intact End, B_6

Only the C_+ and C_0 characteristic lines are relevant at the intact end of the pipeline. Also the flow velocity at the solution point is zero. Consequently, the compatibility Eqs. (21) and (22) coupled with the equation of state are solved iteratively for the unknown values of intact end P, h, ρ, a, and T, with $uj=0$.

Discharge Rate Calculation Algorithm

The discharge rate through the orifice following puncture is dependent on the state of the fluid escaping through it. Depending on the upstream and downstream conditions as well as the composition of the inventory, gas, liquid or a two-phase fluid may be discharged. Fig. 4 is a schematic representation of the pertinent pressures at the orifice junction governing the discharge rate. Pj, is the pressure at the common junction between the two pipeline sections and the orifice. Po and Pd on the other hand are the orifice discharge and downstream pressures, respectively. For choked two-phase release, the discharge pressure, Po is higher than the downstream pressure, Pd. Under such condition, the discharge rate at the puncture is maximum, and no disturbance can be propagated upstream of the puncture. However, under non-critical or no choking conditions, the fluid discharge pressure, Po is equal to the downstream pressure, Pd and the release rate is calculated accordingly.

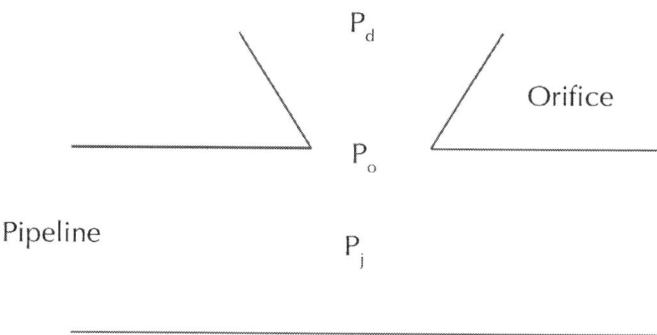

Figure 4: A schematic representation of the pertinent pressures at the orifice junction governing the discharge rate.

The calculations of the choked velocities and hence the subsequent discharge rate for gas or two-phase release requires the application of an energy balance across the orifice. As mentioned above, given the relatively small contact area available for heat transfer and high fluid velocities, we assume that release through the orifice is isentropic. Non-reversibility is accounted for by using a discharge coefficient (Table 1). We further assume that the fluid approaching the orifice is in thermodynamic and phase equilibrium.

Table 1. Hypothetical pipeline case study and Isle of Grain Test Data[a]

Characteristics	Isle of Grain (Test P40)	Condens-able gas
Feed pressure (bara)	21.6	117
Feed temperature (°C)	20	9.85
Pipeline nominal inner diameter (m)	0.154	0.4191
Inlet flow rate (m³/s)	0.0	0.3
Ambient temperature (°C)	19.1	10

Pipe roughness (mm)	0.05	0.05
Ambient pressure (bara)	1.01	1.01
Overall Heat transfer coefficient (W/m²K)	100.0	5.0
Discharge coefficient	1.0	0.8 (assumed)
Time lapsed following cessation of inlet flow (s)	—	90.0
Rupture/puncture diameter (m)	0.150	0.084
Isolated pipeline length (km)	0.1	16
Maximum discretisation time step (s)	Δt 0.9Δt_c	Δt 0.9Δt_c
Pipeline discretisation space step (m)	2.5	250
Puncture distance from high pressure end (km)	0.1	8

Δt_c is the maximum time step calculated using the Courant stability criterion.

a. Condensable gas inventory (molar %): CH_4 (73.6), C_2H_6 (13.4), C_3H_8 (7.4), I-C_4H_{10} (0.4), n-C_4H_{10} (1.0), I-C_5H_{12} (0.08), n-C_5H_{12} (0.07), n-C_6H_{14} (0.02), N_2 (4.03). Isle of Grain inventory (molar %): C_3H_8 (95), n-C_4H_{10} (5).

Fig. 5 is the calculation algorithm for determining the discharge velocity, u0 and hence the discharge flow rate, Qjo through the puncture plane. For a given time step, Δt, the procedure involves an initial guess of the junction pressure, Pj between the two pipeline sections and the orifice. The solution pressure is determined subject to the satisfaction of the appropriate mass and energy balances across the junction control volume (see Fig. 2).

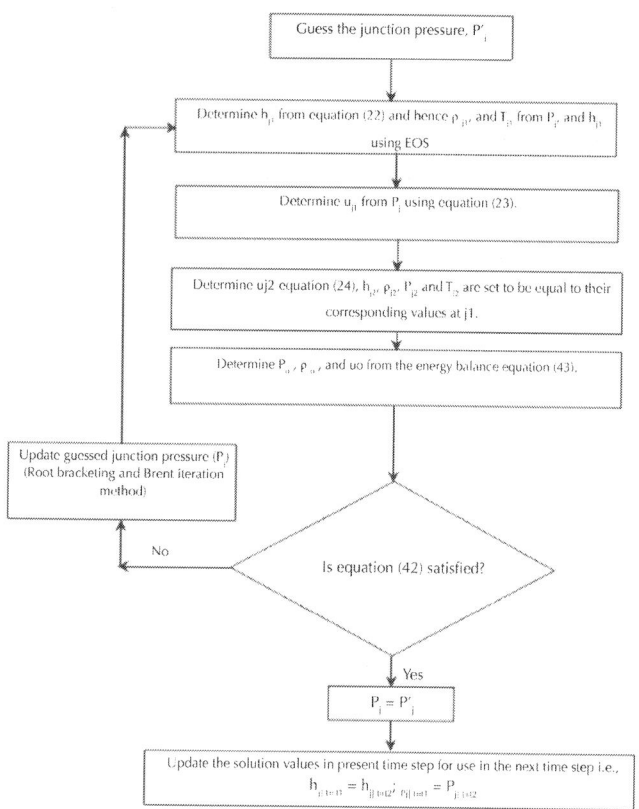

Figure 5: Calculation algorithm for determining discharge rate through orifice following pipeline puncture.

Ignoring the changes in potential energies between the flow approaching and the flow exiting the orifice, the corresponding energy balance across the orifice is given by

$$h_j = h_o + \tfrac{1}{2}u_o^2.$$
(43)

In the case of choked flow, the velocity, u_o in Eq. (43) is replaced by the local speed of sound, a. For single-phase gas flow, a is determined analytically using Eq. (9) in conjunction with the equation of state. In the case of critical two-phase flow, a is determined numerically using Eq. (11).

Under critical flow condition, Eq. (43) is solved iteratively using the Brent iteration method (Press et al., 1992). This involves guessing and updating guessed orifice pressures in conjunction with pressure–entropy (isentropic) flash calculations until Eq. (43) is satisfied. Once a solution is obtained, the flow variables at the orifice (r_o; u_o, T_o, h_o) can be obtained.

RESULTS AND DISCUSSION

The following describes the results of the validation of the above model against the field data obtained from a series of pipeline depressurisation tests conducted by Shell and British Petroleum on the Isle of Grain (Richardson & Saville, 1996). In order to demonstrate the important fluid flow phenomena governing the discharge process, the model is next applied to the hypothetical puncture of a pressurised pipeline containing a condensable hydrocarbon mixture.

Table 1 gives the general characteristics of the two pipelines and prevailing conditions. The Isle of Grain experiments relate to the depressurisation of an instrumented 100 m long and 0.154 m-diameter pipeline containing commercial LPG. In these experiments, the pipeline was punctured following the rupture of a disc placed at one of the closed ends of the pipeline, normal to its major axis. A heat transfer coefficient of 100 w/m^2 is taken, as the pipeline is un-insulated. In the case of the condensable gas simulation however, a more realistic case involving puncture halfway along the length of a partially insulated (heat transfer coefficient of 5 w/m^{2K}) 16Km section of the pipeline isolated by an emergency shut down valve is modelled. For the sake of an example, it is further assumed that inlet flow from a reservoir at the high-pressure end is stopped 90 s after puncture.

The discretisation time, Δt and space, Δx steps employed in the simulation subject to the Courant stability criterion (Courant et al., 1926) are also given.

Validation

Fig. 6, Fig. 7 and Fig. 8 respectively show simulated pressure, temperature and inventory variations with time as compared to the Isle of Grain field data. The state of the fluid during the depressurisation is also indicated in the figures. The pressure and temperature data relate to both the intact (closed) and open ends of the pipeline. As it may be observed, in all cases the model's predictions agree reasonably well with measured data.

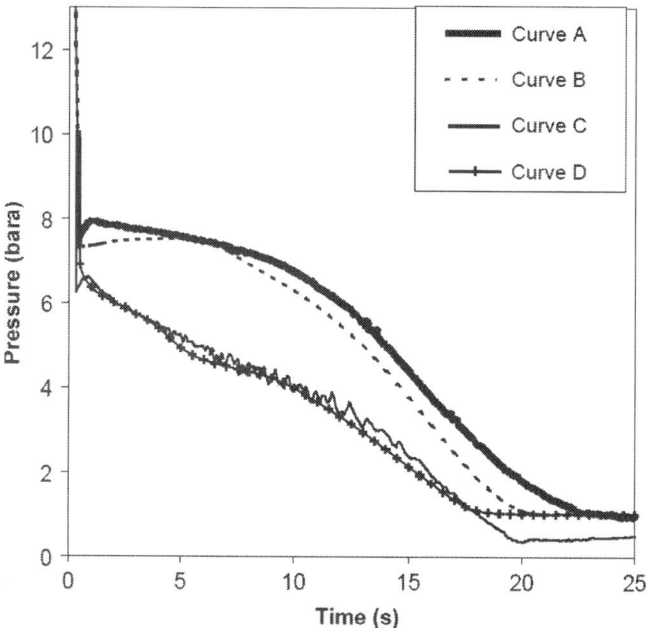

Figure 6. Pressure-time profiles at closed and open ends for the P40 test (LPG). Curve A: field data (closed end); Curve B: model predictions (closed end); Curve C: field data (open end); Curve D: model predictions (open end).

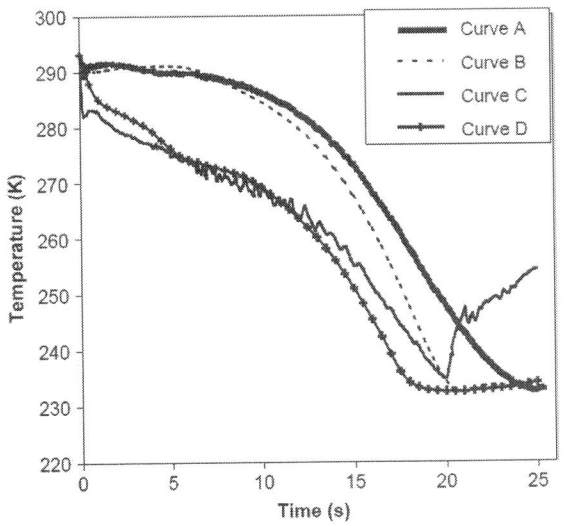

Figure 7. Temperature-time profiles at closed and open ends for the P40 test (LPG). Curve A: field data (closed end); Curve B: model predictions (closed end); Curve C: field data (open end); Curve D: model predictions (open end).

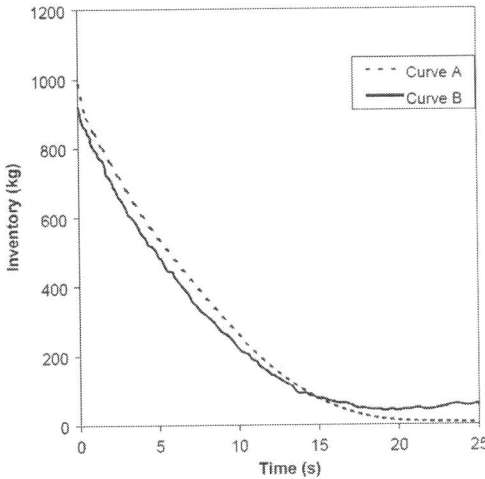

Figure 8. Total line inventory predictions for the P40 test (LPG). Curve A: field data; Curve B: model predictions.

The initial rapid drop in the line pressure (Fig. 6) is due to the transition from liquid to two-phase flow. The observed rapid rise in the measured fluid temperature at the open end (Fig. 7, curve C) towards the end of depressurisation on the other hand corresponds to the transition from two-phase to gas flow and is a consequence of the different fluid/wall heat transfer coefficients. This effect is not seen in the simulated data (Fig. 7, curve D) as a constant heat transfer coefficient is assumed.

Interestingly, the changes in the fluid phase during the depressurisation are not manifested as discontinuities in the variation of inventory with time for both the measured (Fig. 8; curve A) and the simulated data (Fig. 8; curve B).

CONDENSABLE GAS SIMULATIONS

Pressure and Velocity Profiles

Fig. 9 and Fig. 10 respectively show the variations of fluid pressure and velocity along the pipeline at different time intervals following puncture. Referring to Fig. 9, a marked and rapid drop in the line pressure may be observed at the location of the puncture (8km) in the first few seconds (ca. 90s) following puncture. However, this pressure quickly recovers with distance towards the intact end, eventually reaching a value greater than the inlet pressure. The latter is due to the impact of the high velocity flowing fluid with the intact end of the pipeline.

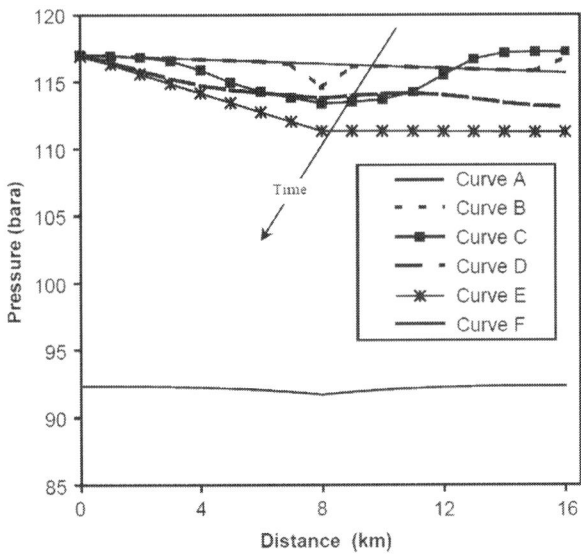

Figure 9: Variation of pressure vs. distance from the high-pressure end of the pipeline at different time intervals following puncture. Curve A: 0s; Curve B: 1s; Curve C: 10s; Curve D: 30s; Curve E: 90s; Curve F: 300s.

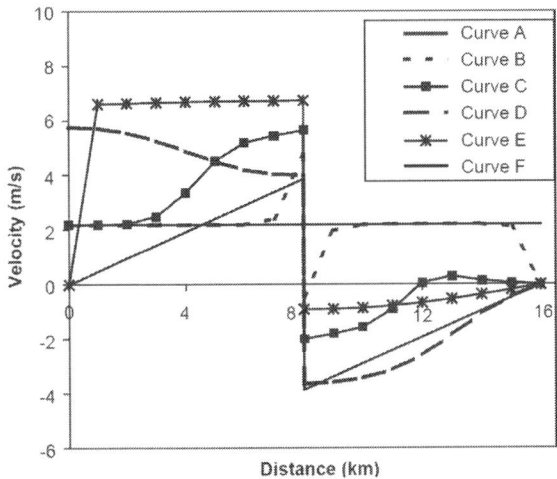

Figure 10: Variation of velocity vs. distance from the high-pressure end of the pipeline at different time intervals following puncture. Curve A: 0s; Curve B: 1s; Curve C: 10s; Curve D: 30s; Curve E: 90s; Curve F: 300s.

The cessation of pumping 90s following puncture results in a rapid and permanent loss of line pressure throughout the pipeline with the effect becoming slightly more pronounced beyond the puncture location. No pressure recovery is encountered in this region. Returning to Fig. 10 showing the line velocity profiles, it may be observed that shortly (1s) after puncture, the flow velocity gradually decreases with distance away from the high-pressure end, but rapidly peaks to a maximum value at the puncture location. This is followed by an instantaneous drop to almost zero velocity before recovering to a value close to the line velocity prior to puncture. The velocity profiles at the subsequent time intervals follow similar trends in the upstream orifice section of the pipeline with the fluid flowing from the high-pressure end towards the puncture. However, downstream of the puncture, a flow reversal is observed where fluid flows from the intact end towards the pump. The cessation of flow is marked by a rather symmetrical but reverse velocity profile at either side of the puncture point.

Fig. 11 gives a schematic representation of the fluid dynamics within the pipeline following puncture. These profiles are generated based on the data given in Fig. 9 and Fig. 10. As it may be observed, upon puncture, part of the high-pressure fluid escapes through the puncture with the remainder flowing towards the intact end of the pipeline. Impact of the fluid with the intact end of the pipeline results in flow reversal and hence the observed fluid flow in both directions in the low-pressure section of the pipeline up to 30s following puncture. Such type of fluid flow behaviour will have a profound effect on the dynamic response of non-return valves used as a means of isolating flow (Mahgerefteh et al., 1997). On the other hand, the accompanying observed pressure surge (see Fig. 9) may undermine the mechanical integrity of the pipeline.

Upon the cessation of pumping, the fluid dynamic profile in the pipeline is similar to that of a vessel discharging through an orifice with no further evidence of flow reversal.

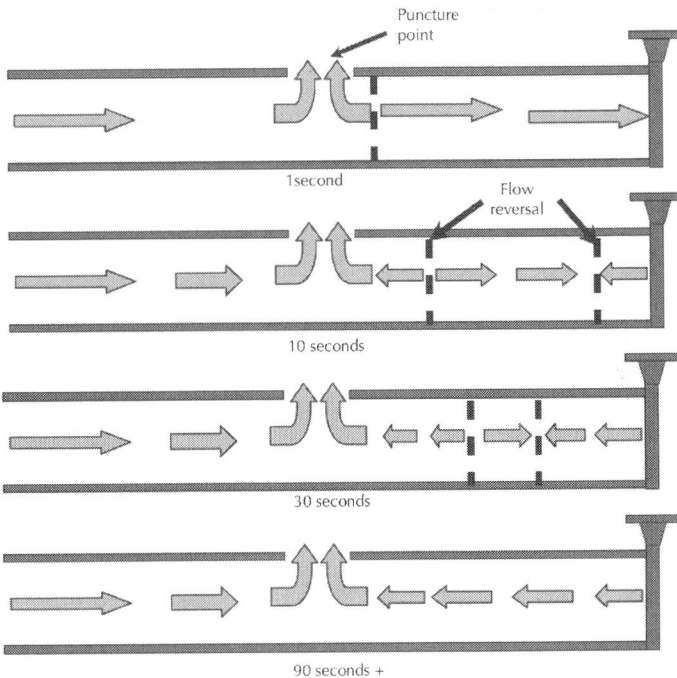

Figure 11: Schematic representation of flow patterns in the pipeline following pipeline puncture.

Discharge Rate

Fig. 12 shows the variation of discharge rate through the puncture with time. As it may be observed, the discharge rate starts at a high value (ca. 125 kg/s) and remaining relatively constant in the first few seconds following puncture. The cessation of feed at 90s following puncture is marked by a gradual reduction in the discharge rate reaching 80kg/s at ca. 800s.

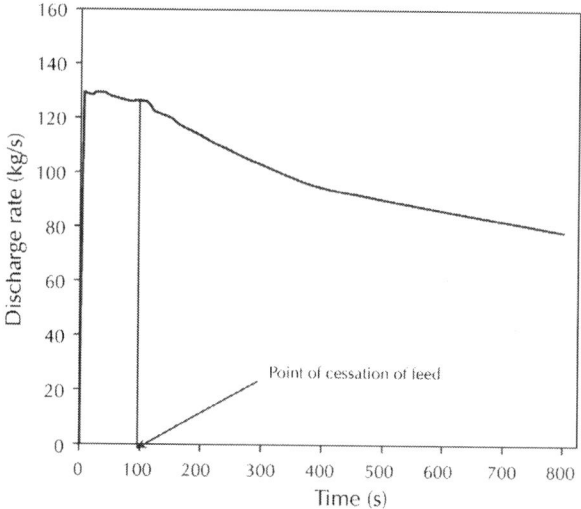

Figure 12: Variation of discharge rate vs. time following pipeline puncture.

CONCLUSIONS

A robust numerical simulation based on the method of characteristics for predicting outflow following pipeline puncture has been developed. Despite the fact that the majority of pipeline accidents involve puncture, most of the models reported so far are based on 1-D axial flow. As such these are restricted to simulating the less likely failure scenario involving full bore pipeline rupture or puncture at one end of the pipeline normal to its major axis. Others simply treat the pipeline as a vessel discharging through an orifice.

Accounting for fluid flow in the radial and the axial directions in the proximity of the puncture, the model described in this paper allows the prediction of the transient release rate through the puncture plane. It also provides detailed insight into some intriguing but important fluid flow phenomena that occur following pipeline puncture. The simulations of the velocity and pressure profiles for example indicate three distinct fluid flow regimes prevailing within the pipeline which govern the variation of the release rate with time. Upon puncture and emergency isolation, part of the high-pressure fluid escapes through the puncture with the remainder continuing to flow towards the low-pressure end of the pipeline. The impact of this high velocity fluid with the intact end of the pipeline causes a pressure surge followed by flow reversal in the downstream pipeline section. Finally, further reduction in the line pressure due to inventory loss results in flow from both of the pipeline sections towards the puncture.

Based on the above findings, it is clear that the simplified release models treating the pipeline as a vessel discharging through an orifice are inappropriate; particularly during the early stages of the depressurisation process when pumping occurs. On the other hand, the observed pressure surge and flow reversal will have important implications in dictating the mechanical integrity of the pipeline and the dynamic response of non-return valves employed for isolating flow following puncture. The model described in this paper is a useful tool in elucidating the important fluid dynamic phenomena governing the discharge rate following the puncture of pressurised pipelines. The specially developed solution technique based on pressure–enthalpy flash results in a significant reduction in computing run times. This development addresses a major difficulty synonymous with the numerical simulation of pipeline ruptures.

ACKNOWLEDGMENTS

The authors are grateful to the NATO-Russia Collaborative Linkage programme for providing funding in relation to the above work.

REFERENCES

1.   Assael, M. J., Martin Trusler, J. P., & Tsolakis, T. (1996). Thermophysical properties of >uids. London: Imperial College Press.

2. Barlas, S. (1999). Pipeline safety law reauthorisation. Pipeline & Gas Journal 226, 6.

3. Bendiksen, K. H., Malnes, D., Moe, R., & Nul, S. (1991). The dynamic two-fluid model OLGA: Theory and applications. SPE Production Engineering, 6, 171.

4. Bisgaard, C., Sorensen, H., & Spangenberg, S. (1987). A finite element method for transient compressible flow in pipelines. International Journal for Numerical Methods in Fluids, 7, 291.

5. Bond, J. (2002). Institute of Chemical Engineers accidents database. Rugby, UK: Institute of Chemical Engineers.

6. Chen, J. R., Richardson, S. M., & Saville, G. (1992). Numerical simulation of full-bore ruptures of pipelines containing perfect gases. Transactions of the Institute of Chemical Engineers, Part B, 70, 59.

7. Chen, J. R., Richardson, S. M., & Saville, G. (1993). A simplified numerical method for transient two-phase pipe flow. Transactions of the Institute of Chemical Engineers, 71A, 304.

8. Chen, J. R., Richardson, S. M., & Saville, G. (1995a). Modelling of two-phase blowdown from pipelines—I. a hyperbolic model based on variational principles. Chemical Engineering Science, 50, 695.

9. Chen, J. R., Richardson, S. M., & Saville, G. (1995b). Modelling of two-phase blowdown from pipelines—ii. a simplified numerical method for multi-component mixtures. Chemical Engineering Science, 50, 2173.

10. Courant, R., Friedrichs, K. O., & Lewy, H. (1926). Uber die Partiellen Differential gleichungen der Mathematischen, Physik. Mathematische Annalen, 100, 32.

11. Cullen, W. D. (1990). The public inquiry into the Piper Alpha Disaster. London: Department of Energy, HMSO.

12. de Reuck, K. M., Craven, R. J. B., & Elhassan, A. E. (1996). In J. Millat, J. H. Dymond, & C. A. Nieto de Castro (Eds.), Transport properties of fluids: Their correlation, prediction

and estimation. IUPAC, Cambridge: Cambridge University Press.

13. Flatt, R. (1986). Unsteady compressible flow in long pipelines following a rupture. International Journal for Numerical Methods in Fluids, 6, 83.

14. Fletcher, S. (2001a). US senate ready to act on pipeline safety. Oil & Gas Journal, 99(6), 58.

15. Fletcher, S. (2001b). Pipeline safety rules may stretch. Oil & Gas Journal, 99(7), 58.

16. Lang, E. (1991). Gas flow in pipelines following a rupture computed by a spectral method. Journal of Applied Mathematics and Physics (ZAMP), 42, 183.

17. Lydell, B. O. Y. (2000). Pipe failure probability-The Thomas paper revisited. Reliability Engineering and System Safety, 68, 207.

18. Mahgerefteh, H., Saha, P., & Economou, I. G. (1997). A study of the dynamic response of emergency shutdown valves following full bore rupture of gas pipelines. Transactions of the Institute of Chemical Engineers Part B, 75, 201.

19. Mahgerefteh, H., Saha, P., & Economou, I. G. (1999). Fast numerical simulation for full bore rupture of pressurized pipelines. A.I.Ch.E. Journal, 45(6), 1191.

20. Mahgerefteh, H., Saha, P., & Economou, I. G. (2000). Modelling fluid phase transition effects on dynamic behaviour of ESDV. A.I.Ch.E. Journal, 46(5), 997.

21. Massey, B. S. (1983). Mechanics of fluids. Wokingham: Van No strand Reinhold.

22. Michelsen, M. L. (1982a). The isothermal #ash problem. Part I. Stability. Fluid Phase Equilibria, 9, 1.

23. Michelsen, M. L. (1982b). The isothermal flash problem. Part II. Phase-splitcalculation. Fluid Phase Equilibria, 9, 21.

24. Michelsen, M. L. (1987). Multi-phase isenthalpic and isentropic flash algorithms. Fluid Phase Equilibria, 33, 13.

25. Montiel, H., Vilchez, J. A., Casal, J., & Arnaldos, J. (1998). Mathematical modelling of accidental gas releases. Journal of Hazardous Materials, 59, 211.

26. Olorunmaiye, J. A., & Imide, N. E. (1993). Computation of natural gas pipeline rupture problems using the method of characteristics. Journal of Hazardous Materials, 34, 81.

27. Peng, D. Y., & Robinson, D. B. (1976). A new two-constant equation of state. Industrial and Engineering Chemistry Fundamentals, 15, 59.

28. Press, W. H., Teukolsky, S. A., Vetterling, W. T., & Flannery, B. P. (1992). Numerical recipes in FORTRAN 77: The art of scientiCc computing (2nd ed.). Cambridge: Cambridge University Press.

29. Richardson, S. M., & Saville, G. (1996). Blowdown of LPG pipelines. Transactions of the Institute of Chemical Engineers, 74B, 236.

30. Shoup, G., Xiao J. J., & Romma, J. O. (1998). Multiphase pipeline blowdown simulation and comparison to field data. First North American conference on multiphase technology, BHR Group Conference Series No. 31, Banff, Canada, Vol. 3, June 10–11.

31. True, W. R. (2001). Regulatory actions loom for US pipelines in 2001. Oil & Gas Journal, 99.1, 70.

32. Versteeg, H. K., & Malalasekera, W. (1995). An introduction to computational fluid dynamics: The Cnite volume method (p. 10). London: Prentice-Hall.

33. Woodward, J. L., & Mudan, K. S. (1991). Liquid and gas discharge rates through holes in process vessels. Journal of Loss Prevention in the Process Industries, 4, 161.

34. Wylie, E. B., & Streeter, V. L. (1993). Fluid transients in systems. Englewood Cli?s, NJ: Prentice-Hall, 51.

35. Young-Do, J., & Bum, J. A. (2003). A simple model for the release rate of hazardous gas from a hole on high-pressure pipelines. Journal of Hazardous Materials, A97, 31.

36. Zhou, X. X., Lea, C. J., Bilo, M., & Maddison, T. E. (1997).
     Three-dimensional computational fluid dynamic modelling
     of natural gas releases from high-pressure pipelines. Pipes
     and Pipelines International, 42(5) 13.

37. Zucrow, M. J., & Hoffman, J. D. (1976). Gas dynamics, Vols. I
     and II. 297. New York: Wiley

Citations

CHAPTER 1

M. Araújo, S. Neto and A. Lima, "Theoretical Evaluation of Two-Phase Flow in a Horizontal Duct with Leaks,"Advances in Chemical Engineering and Science, Vol. 3 No. 4A, 2013, pp. 6-14. doi: 10.4236/aces.2013.34A1002.

CHAPTER 2

Tianmin Wang, Hiroshi Mori, Chong Zhang, Ken Kurokawa, Xin-Hui Xing, and Takuji Yamada, DomSign: a top-down annotation pipeline to enlarge enzyme space in the protein universe, doi: 10.1186/s12859-015-0499-y.

CHAPTER 3

S.K. Lahiri, K.C. Ghanta, Development of an artificial neural network correlation for prediction of hold-up of slurry transport in pipelines, Chemical Engineering Science, Volume 63, Issue 6, March 2008, Pages 1497-1509, ISSN 0009-2509, http://dx.doi.org/10.1016/j.ces.2007.11.030.

CHAPTER 4

Dmitry Eskin, Applicability of a Taylor–Couette device to characterization of turbulent drag reduction in a pipeline, doi:10.1016/j.ces.2014.05.016.

CHAPTER 5

Haroun Mahgerefteh, Solomon Brown, Garfield Denton, Modelling the impact of stream impurities on ductile fractures in CO_2 pipelines, Chemical Engineering Science, Volume 74, 28 May 2012, Pages 200-210, ISSN 0009-2509, http://dx.doi.org/10.1016/j.ces.2012.02.037.

CHAPTER 6

S.L Ke, H.C Ti, Transient analysis of isothermal gas flow in pipeline network, Chemical Engineering Journal, Volume 76, Issue 2, February 2000, Pages 169-177, ISSN 1385-8947, http://dx.doi.org/10.1016/S1385-8947(99)00122-9.

CHAPTER 7

B.V. Balakin, A.C. Hoffmann, P. Kosinski, S. Høiland, Turbulent flow of hydrates in a pipeline of complex configuration, Chemical Engineering Science, Volume 65, Issue 17, 1 September 2010, Pages 5007-5017, ISSN 0009-2509, http://dx.doi.org/10.1016/j.ces.2010.06.005.

CHAPTER 8

Adeyemi Oke, Haroun Mahgerefteh, Ioannis Economou, Yuri Rykov, A transient outflow model for pipeline puncture, Chemical Engineering Science, Volume 58, Issue 20, October 2003, Pages 4591-4604, ISSN 0009-2509, http://dx.doi.org/10.1016/S0009-2509(03)00338-5.

Index